PRACTICAL ISOTOPE HYDROLOGY

NIPA® GENX ELECTRONIC RESOURCES & SOLUTIONS P. LTD.
New Delhi-110 034

www.nipaers.com
Browse, Search, Read & Buy...
Print Books
eChapters
eBooks
Publishers
Forthcoming
Subject Catalogues
New Titles
Online Resources on:
✓ Current Affairs
✓ Reasoning, Logic & Aptitude
✓ Competitive Examinations
✓ English Language Lab
✓ Writing & Pronunciation Tools
✓ Personality Development
✓ Online Programmes for Professionals
✓ Interview Preparation
Pay Using
PhonePe
Google Pay
Paytm
PayPal
UPI
CC Avenue

PRACTICAL ISOTOPE HYDROLOGY

S. M. Rao
Associate Director (Retd.), Isotope Group
Bhabha Atomic Research Centre
Mumbai
and
Presently, Member (Research Council)
Centre for Water Resources Development and Management
Kozhikode

NIPA® GENX ELECTRONIC RESOURCES & SOLUTIONS P. LTD.
New Delhi-110 034

NIPA® GENX ELECTRONIC
RESOURCES & SOLUTIONS P. LTD.

101,103, Vikas Surya Plaza, CU Block
L.S.C. Market, Pitam Pura, New Delhi-110 034
Ph : +91 11 27341616, 27341717, 27341718
E-mail: newindiapublishingagency@gmail.com
web: www.nipabooks.com

For customer assistance, please contact
Phone: + 91-11-27 34 17 17
Fax: + 91-11- 27 34 16 16
E-Mail: feedbacks@nipabooks.com

ISBN: 978-81-19254-50-7

Composed and Designed by NIPA®.

Dedicated

to the

BHABHA ATOMIC RESEARCH CENTRE

PREFACE

It has been of some concern that in spite of over four decades of development and demonstration, isotope hydrology is yet to find its rightful place in field practice in India and in many other countries. Isotope hydrology is still not taught as part of hydrology syllabus in many universities. It is probably not surprising when you notice the level of application of geochemistry in groundwater investigations. In real practice, chemistry is generally restricted to determinations of water quality, more specifically to potability. Geochemistry, particularly trace element chemistry, is rarely used as a tool to study groundwater evolution and residence. Isotope hydrology, being an extension of geochemistry, cannot possibly claim a better status. This is particularly unfortunate in view of India's ever-increasing need for utilizable water, particularly groundwater. Nearly 60% of India's groundwater occurs in hard rocks. The field hydrologist knows how frustrating it is to depend on methods developed for characterizing alluvial aquifers. India's hydrological diversity is immense from glaciers to humid tropics to deserts with attendant huge range of water resources problems. With our long-standing experience in the management of water resources, it should be evident India needs to use all available tools. Isotope hydrochemistry is one such tool.

Isotope hydrology is definitely not a panacea for all hydrological ills. It has value when integrated with the other well-established tools like geology, hydrogeology and geophysics and, of course geochemistry and remote sensing.

It is tempting to compare the level of absorption of scientific developments in medical practice with the level in hydrological practice. The speed with which the medical establishments in India adapted to computed tomography, magnetic resonance imaging, nuclear medicine and radiation sterilization of medical products is commendable. The same cannot be said about the water resources scene in India. Computer modeling of hydrology data and remote sensing are the exceptions. Here again, diversity and the quality of data available for modeling need to be strengthened and modernized.

If one browses through refereed international journals on water resources, the scientific talent in the country would be evident. India has also accumulated significant experience in isotope hydrology, but not wide enough to have a regular say in the country's water resources scene. There are a few institutions with facilities for isotope work. Some of them are the Bhabha Atomic Research Centre, Mumbai; Physical Research Laboratory, Ahmedabad; National Geophysical Research Institute, Hyderabad; National Institute of Hydrology, Roorkee; Indian Agricultural Research Institute, Delhi and Centre for Water Resources Development and Management, Kozhikode. But, oganisations like the Central Ground Water Board who are directly responsible for water resources development in India are still to have in-house facilities for isotope hydrology.

Then what are the constraints? There is a certain lack of stress on basic sciences in hydrology education. Geochemistry, as mentioned earlier, is not fully exploited as a tool in groundwater studies. Isotope hydrology, being closely associated with nuclear sciences, may not be easily usable by a field hydrologist who had no earlier training. Field hydrologists are rarely prepared to use techniques not taught in the university. Compared to conventional equipment and instruments used in hydrology investigations, isotope analytical equipment like mass spectrometers and scintillation counters are relatively more expensive. But, when compared with our national budget for water resources development, the benefit to cost ratio would make the expense negligible. Decision makers need to be properly briefed on these aspects.

Some excellent books are available, in literature, on isotope hydrology. They include

1. Handbook of Environmental Isotope Geochemistry (Ed. P. Fritz, J. Ch. Fontes)
2. Environmental Isotopes in Hydrogeology (I. Clark, P.Fritz)
3. Isotopes in Catchment Hydrology (Ed. C. Kendall, J.J. McDonnell)
4. Chemical and Isotopic Groundwater Hydrology - The Applied Approach (E. Mazor)
5. Environmental Isotopes in Hydrological Cycle, Principles and Applications (Ed. W.G. Mook), UNESCO/IAE Series, 6 volumes.

Research scientists with deep insight into the scientific aspects of isotope methodologies wrote most of the books. These books are of immense value to research hydrologists who had already gained some insight into the basic science aspects of isotope hydrology. The books do not appear to be readily usable by practicing hydrologists who may be looking for solutions to field problems. In this book an attempt is made to simplify the basic science concepts and focus on field applications.

Hydrology is region-specific. It is always the local hydrologist who has a better feel of the hydrology of a particular region and a better understanding of its problems. He/she may need better tools. Hence an attempt is made in this book to give as many Indian examples as possible. The primary objective is to provide information on the capabilities of isotope hydrology and encourage the hydrologist to seek possible isotope solutions to field problems. This book may hopefully help to introduce isotope hydrology as part of curriculum in Indian universities. This book in its present form cannot claim to be a textbook for such a course, but may have a role in combination with the other books mentioned above.

The book is divided into an introduction as to what isotope hydrology is all about and why should isotope methods be used. Then follows the basis and relevance of various isotope methods and finally field applications in various stages of the hydrological cycle.

I acknowledge with thanks that the IAEA/UNESCO publication Environmental Isotopes in Hydrological Cycle, Principles and Applications (Ed. W.G. Mook), 6 volumes have been particularly useful in many ways.

Mumbai **S.M. Rao**
March, 2006

CONTENTS

1 INTRODUCTION

"Hydrology is the science that treats the waters of the earth, their occurrence, circulation and distribution, their chemical and physical properties, and their reaction with their environment, including their relation with living things. The domain of hydrology embraces the full life history of water on the earth." (Federal Council for Science and Technology, "Scientific Hydrology", Washington, 1962)

This definition puts in a nutshell what is most often inadequately done before undertaking exploitation of water resources. In short, it is the scientific inputs that are likely to be missed while planning water resources management projects. It is the "full life history of water" part of the above definition that is of specific relevance to the theme of this book, Isotope Hydrology.

Conventional hydrology is strong on measurement and quantification of different parts of the hydrological cycle (Fig.1.1), be it precipitation, run-off, evaporation, transpiration, infiltration, groundwater dynamics etc. Through chemistry and isotopes, we try to understand the processes themselves in order to delineate the history of that part of the hydrological cycle that is being studied. For example, if we are measuring river flow, we would like to know the contribution of different components; base flow, rainfall run-off and snow melt, if any.

Let's see what we know and what do not know in the conventional hydrological practice. We know surface waters evaporate, form into clouds and precipitation occurs under favourable weather conditions. What we often do not know is the source of moisture, oceanic or inland.

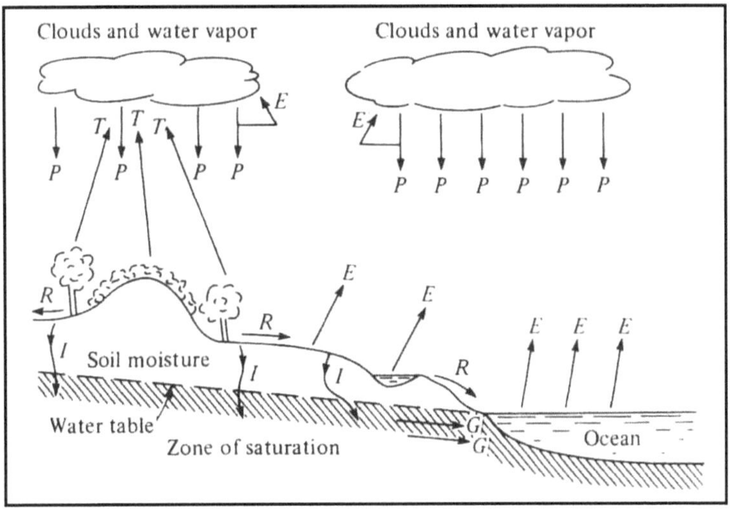

Figure 1.1 Hydrological cycle, T: Transpiration, E: Evaporation, P: Precipitation, R: Surface runoff, G: Groundwater flow, I: Infiltration

As mentioned above, we can measure river flow, but component-wise differentiation of base flow, snow melt and rainfall run off is not easy. It is possible to estimate recharge to unconfined aquifers, but here again, we need to know how much of it is from direct rainfall and from other surface water resources like reservoirs, canals and rivers and also what is the time delay in water reaching the groundwater table from the various sources. Deep and confined aquifers are becoming increasingly important in water resources development and management. Our understanding of the processes of recharge to these aquifers is not adequate enough for their optimum utilization. Seawater intrusion into coastal aquifers can be identified, but what is not immediately clear is whether it is a palaeo-seawater entrenched in foreshore sediments or it is present day seawater.

Isotope Hydrology aims to help fill in the above-mentioned gaps in our understanding of the hydrological cycle, particularly on a regional scale to enable better management of water resources. It should be emphasized that isotope hydrology is a supportive science. Best results are achieved only when integrated with well-established methods in surface hydrology, hydraulics, hydrogeology, geophysics and finally geochemistry.

Isotope hydrology is part of tracer hydrology. What then is tracer hydrology?

1.1 Tracer Hydrology

In studies on dynamics of any system, be it biological, industrial or hydrological, it is always rewarding to identify a representative part of the system or a group of molecules and follow its behaviour as it progresses through the system. The obvious problem is how to differentiate that part or group of molecules from the rest of the system. It is here the tracer concept helps.

Tracer is either an intrinsic component of the system or artificially introduced into the system and whose physical and chemical variations are interpretable with reference to the system dynamics.

Blood glucose in human body is an example of an intrinsic tracer. So also, as we shall see later, are the natural isotopic compositions of water and of the dissolved salts.

Radio-opaque dye used to study kidney function is an example of an artificially injected tracer. So also are common salt, organic dyes like rhodamines or radioactive bromide used to study hydrological systems. Choice of an injected tracer is rather delicate, as it should behave like the traced material (water in our case) and also distinguishable for detection. A classical example of application in hydrology is short-range measurement of groundwater velocities using common salt.

Broadly speaking, the kinetic behaviour of any liquid water tracer, both intrinsic and injected, should be identical or nearly identical to water in a given environment (river, lake or aquifer) except during phase transitions. While moving close to the velocity of traced water molecules (no observable density currents) an injected tracer should have neither sources or sinks. The problem of "sources" is usually taken care of by measurement of any 'background' concentration and its variation. By 'background', we mean the levels naturally present in the system or remnants of a previous investigation. In the classical example of common salt as the tracer, the chloride level naturally present in water is the 'background'.

It should be ensured the tracer is not removed or retarded by such physico-chemical processes as adsorption, ion exchange, precipitation, photochemical degradation and biogeochemical influences. Some of these processes are unavoidable, particularly in the case of intrinsic tracers like dissolved salts (Ex: bacterial reduction of sulphate to sulphide in groundwater). In such cases, the tracer hydrologist is called upon to provide right interpretation.

An additional requirement for an injected tracer is it should be detectable at the maximum dilution envisaged.

Tracer hydrology lays great stress on the selection of a suitable tracer for a given application. A host of tracer materials ranging from simple salts like sodium chloride and potassium iodide to organic dyes like sulpharhodamine G, rhodamine WT and uranine [1] have been used with varying degrees of success for over a century. Among electrolytes, anions and metal complexes have distinct advantages, particularly in groundwater studies. Since soils are usually negatively charged, anionic tracers are not easily adsorbed or retarded vis-à-vis water. Metal complexes have high stability constants. With the advent of radioactive isotope tracers which can be produced in anionic and metal complex forms and which are detectable in extremely high dilutions, injected tracer hydrology is greatly benefited [2]. Most of these applications are for short-range, but not necessarily short term studies. The injected isotope tracers are today largely applied in studying seepages in dams, rainfall recharge measurement to unconfined aquifers and in special cases of stream flow measurements.

The level of application of intrinsic isotope tracers in hydrology is far higher. The applications include

- Component-wise separation of stream flow; base flow, run-off and snowmelt
- Studies on lake dynamics including estimation of retention time and residence time of inflowing groundwater
- Identification of existence or absence of recharge and understanding of recharge processes
- Studies on surface water- aquifer and aquifer- aquifer interconnections

- Studies on groundwater salinization processes including seawater intrusion into coastal aquifers.
- Age dating of old ground waters
- Arid Zone Hydrology including studies on buried channels
- Studies on groundwater pollution, mainly from natural causes (Ex: arsenic, nitrate etc.)

1.2 What are Isotopes?

It is common knowledge that all elements are composed of atoms and each atom has a central nucleus consisting of positively charged protons and electrically neutral neutrons. Negatively charged electrons orbit the nucleus. To keep the whole atom electrically neutral, the number of protons in the nucleus equals the number of electrons orbiting it.

The identity of an atomic element and its chemical behaviour depend on the number of protons/electrons in it. Hydrogen has 1 proton, oxygen 8, calcium 20, silver 47, gold 79 and uranium 92. This number tells us why oxygen is unlike gold, which is unlike uranium. The proton number of any element is called the Atomic Number of that element. Let's call it 'Z'.

What about the neutrons? In most of the elements of common interest, the number of neutrons is more or less the same as the proton number, but not exactly. This number has, however, no effect on the chemistry of the atomic element. In fact all elements occur in nature with variable neutron number. The total number of nucleons (protons + neutrons) in the nucleus of an atomic element is called the Mass Number of that element. Let's call it A.

It is now clear that for a given element, the atomic number Z is invariant whereas the mass number A is variable. Here comes the concept of ***isotopes***. The species of an atomic element with varying mass number are called the *isotopes* of that element. The isotope of any element X is represented by

$$ {}^{A}_{Z}X \quad \text{or simply as} \quad {}^{A}X \quad (\text{Ex: } {}^{3}H, {}^{14}C, {}^{198}Au) $$

Often isotopes of elements are also written as Hydrogen-3, Carbon-14, Gold-198 etc., or even more simply as H-3, C-14 and Au-198.

1.2.1 Isotopes and their Characteristics

Isotopes are either naturally occurring or manmade; the latter in a nuclear reactor or in a particle accelerator. Secondly they are either stable or radioactive. Whereas both stable and radioisotopes occur in nature, manmade isotopes are all unstable/radioactive.

The number of isotopes naturally occurring for each element varies widely from 0 to 10 (0 is for Technetium, which is not known to occur naturally). These are usually stable (non-radioactive) for light elements up to bismuth with exceptions like Potassium-40 and Rubidium-87, which are naturally occurring radioisotopes of light elements. These are primordial radioactive isotopes i.e. those formed in the building up of terrestrial matter and are still present today. There is yet another set of radioisotopes which are not primordial, but produced by cosmic ray interaction with air in the upper atmosphere (Ex: Tritium or Hydrogen-3 and Cabon-14) and also by terrestrial radiation (neutrons) in the earth's crust (Ex: Chlorine-36)

All isotopes of elements heavier than bismuth, naturally occurring or otherwise, are all radioactive. The naturally occurring radioisotopes in the trans-bismuth range belong to what are known as the "Natural Radioactive Families" of Thorum-232, Uranium-235 and Uranium-238 (Fig.1.2).

Radioactivity

Radioactivity is a process by which an unstable nucleus, which has excess energy, tries to get back to a stable state through emission of some form of radiation, either electromagnetic or particulate. This is not unlike a hot body giving off thermal radiation to reach the temperature of its environment.

But what distinguishes these radiations from other forms of radiations such as visible light and heat? Radioisotope radiations are ionizing radiations. By virtue of their higher energy or electrical charge in the case of particulates, their interaction with matter causes ionization/ excitation of the atoms. Ionization results in the removal of one or more orbital electrons from a neutral atom rendering it positively charged whereas excitation merely raises the orbital electron to a higher energy level keeping the atom still electrically neutral.

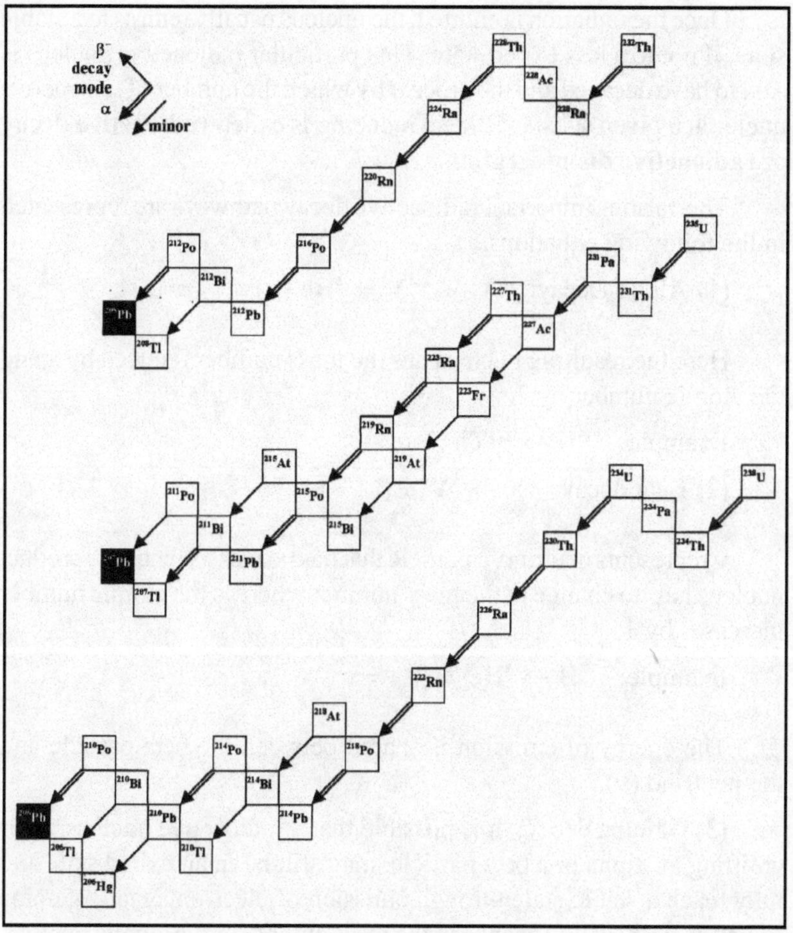

Figure.1.2 Natural Radioactive Families of Thorium, Actinium and Uranium. Black box represents the end product, a stable isotope of lead in all cases

The radiation emitted by a radioactive nucleus may either be particulate or electromagnetic or both. Particulate radiations include alpha rays (α, same as ^{4}He nuclei), beta particles (β^-, same as electrons) and neutrons. Electromagnetic radiation includes x-rays and gamma (γ) rays (very high energy electromagnetic radiation similar to x-rays). The different nomenclature of beta and gamma, which emanate from within the nucleus is to distinguish them from electrons and x-rays, which are of extra-nuclear origin.

Once the radiation is emitted, the nucleus usually returns to a stable state, if not to a less exited state. That particular radioactive nucleus is said to have decayed and the process by which the number of radioactive nuclei in a given substance keep reducing is called **radioactive decay** or **radioactive disintegration.**

The various important radioactive decay pathways are represented in the following equations.

(1) Alpha Decay: $^{A}\mathbf{X} \rightarrow {}^{A-4}\mathbf{Y} + {}^{4}\mathbf{He}\ (\alpha)$

Here the resulting nucleus has the mass number reduced by 4 and the atomic number by 2.

Example: $^{235}\mathbf{U} \rightarrow {}^{231}\mathbf{Th} + \alpha$

(2) Beta decay: $^{A}\mathbf{X} \rightarrow {}^{A}\mathbf{Y} + \beta^{-} + \nu$

ν represents neutrino, a particle that has no mass. Hence the product nucleus has no change in the mass number whereas the atomic number increases by 1.

Example: $^{3}\mathbf{H} \rightarrow {}^{3}\mathbf{He} + \beta^{-} + \nu$

The energy of emission is shared between the beta particle and the neutrino (ν).

(3) Gamma decay: It is possible that a radioactive nucleus, after emitting an alpha or a beta particle, may still be in an exited state and may reach a stable state through emission of electromagnetic gamma radiation without any change in the mass number or the atomic number, that is to say the isotope remains the same before and after gamma emission.

Example: $^{60}\mathbf{Co} \rightarrow {}^{60}\mathbf{Ni}^{*} + \beta^{-} + \nu$

$^{60}\mathbf{Ni}^{*} \rightarrow {}^{60}\mathbf{Ni} + \gamma$

Radioactive Decay Process

Emission of radioactive radiations is a random process. It is not possible to say which particular nucleus would decay at what time, but one can predict with reasonable certainty how many nuclei of a given radioisotope would undergo disintegration in a given time. Like in similar stochastic processes, the number of disintegrations per unit time is proportional to the total number **N** of radioactive nuclei present in a given quantity of radioactive material, i.e.

$$\mathbf{dN/dt = N\lambda} \qquad (1.1)$$

where λ is called the decay constant.

As can be deduced from the above, radioactivity is quantified by the number of disintegrations per second (dps) irrespective of the type, number and energies of radiations emitted in each disintegration. The unit of radioactivity is ***Becquerel* (Bq*)***

$$\mathbf{1Bq = 1dps}$$

The older unit Curie ($1Ci = 3.7x10^{10}$ dps = 37 GBq) is still found to be convenient by many users.

Half-life

As mentioned above, the radioactive decay is a continuous and natural process and the quantity of radioactivity (dps) in a given substance continues to decrease in an exponential way. From Eq 1.1 above, we can get

$$\mathbf{N = N_0 e^{-\lambda t}} \qquad (1.2)$$

Where (N_0 – N) nuclei decayed in time t.

Half-life **($T_{1/2}$)** is defined as the time required for the radioactivity to reduce to half its initial value i.e. when $N/N_0 = 1/2$. From Eq 1.2, we can show that

$$\mathbf{Half\text{-}life\ T_{1/2} = ln2/\lambda = 0.693/\lambda} \qquad (1.3)$$

Energies of radiation

The energies of radioactive radiations are expressed in electron volts **(eV).** 1 eV is the energy gained by an electron while falling through a potential difference of 1 volt.

$$\mathbf{1\ eV = 1.6x10^{-12}\ ergs}$$

The energies of radioactive radiations are normally in the ranges of **keV** (10^3 eV) and **MeV** (10^6 eV). It is important to remember that whereas alpha (α) and gamma (γ) rays are emitted with discrete energies, irrespective of the number of rays emitted per disintegration, the beta (β) rays exhibit a continuous spectrum of energies up to a characteristic maximum. It is this maximum energy, which is traditionally quoted for a beta-emitting isotope.

1.2.2 How are Radioisotopes Produced?

Specially prepared targets are placed in suitable and specially designated positions in nuclear reactors. When bombarded with fast or slow neutrons, one or more of several possible nuclear reactions take place leading to the radioactivation of the target. These include **(n,γ)**, **(n,p)** and **(n,α)** reactions. In each bracket, the first letter 'n' refers to the bombarding neutron and the second to the radiation spontaneously emitted from the target nucleus during the reaction (γ: gamma, p: proton, α: alpha)

Some examples of neutron activation reactions leading to production of isotopes of hydrological interest are as follows.

(n,γ) processes

$$^{59}Co\ (n,\gamma)\ ^{60}Co$$

$$^{81}Br\ (n,\gamma)\ ^{82}Br$$

Some (n,γ) processes are followed by beta decay leading to the production of a desired isotope. An important example is

$$^{130}Te\ (n,\gamma)\ ^{131}Te \rightarrow (\beta^-) \rightarrow\ ^{131}I$$

One important advantage in the above process is that the product radioisotope can be easily separated from the target, as they are two different elements.

(n,p) and (n,α) processes

Though some of the (n, p) and (n,α) reactions are feasible with slow neutrons, many need fast neutrons as the emitted particles are charged. Examples of production of isotopes of interest to hydrology are

$$^{6}Li\ (n,\alpha)\ ^{3}H$$

$$^{14}N(n,p)\ ^{14}C$$

Radioisotopes are later processed in special radiological laboratories to produce tracers of desired chemical composition. Examples are bromine-82 as $\mathbf{NH_4\,^{82}Br}$, cobalt-60 as $\mathbf{K_3\,^{60}Co\,(CN)_6}$, iodine-131 as $\mathbf{K\,^{131}I}$, gold-198 as chloroauric acid ($\mathbf{HAuCl_4}$) and tritium (hydrogen-3) as tritiated water ($\mathbf{^1H^3HO}$).

As mentioned earlier, there is a class of radioisotopes continuously produced in the environment. It is either by the interaction of cosmic rays with air components in the upper atmosphere or by the interaction of secondary cosmic rays reaching the earth's surface. The terrestrial radiogenic neutrons and alpha particles (from the natural uranium and thorium decay series, see Fig.1.2) also interact with soil and rock producing radioisotopes. Among them, Tritium or ^3H (12.32 a), ^{14}C (5730 a) and ^{36}Cl (301 ka) (a = years) are of particular interest for ground water dating and evaluation.

Cosmic ray neutrons produce tritium (^3H) and carbon-14 in the upper atmosphere through the following reactions.

$$\mathbf{^{14}N\ (n,\ ^3H)\ ^{12}C}$$

$$\mathbf{^{14}N\ (n,\ p)\ ^{14}C}$$

They are both quickly oxidized to tritiated water and CO_2 respectively and enter the hydrological cycle.

These radioisotopes were also produced during the atmospheric testing of thermonuclear weapons by the USA and the erstwhile Soviet Union in the late fifties and early sixties before the test ban treaty in 1963 and entered the hydrological cycle. Special attention is paid to this aspect while interpreting the relevant radioisotope data.

I.2.3 Radioisotopes of Hydrological Interest

Following is a summary of important radioisotopes used in hydrology investigations either as intrinsic or injected tracers.

Hydrogen-3 or tritium, a beta emitter (Max.Energy or E_{max}: 0.018 MeV) with a half- life of 12.32 years (In water it is present as part of the water molecule. As an injected tracer it is mainly used to trace soil moisture transport in the unsaturated zone and as environmental tracer to age-date younger groundwater and to study groundwater dynamics).

Carbon-14, a beta emitter (E_{max}: 0.156 MeV) with a half-life of 5730 years (It is present in water as carbonate and bicarbonate. As an environmental tracer it is mainly used to age date old ground waters)

Chlorine-36, a beta emitter (E_{max}: ~ 0.7 MeV) with a half-life of 300,000 years. (As an environmental tracer, it is present in water as chloride. It is used to age date extremely old groundwater, but only rarely due to difficulties in assay and interpretation)

Cobalt-60, a beta-gamma emitter (Gamma energies: 1.17 & 1.33 MeV) with a half- life of 5.27 years. (Injected tracer as $K_3\,^{60}Co(CN)_6$ complex, rarely to trace soil moisture transport in the unsaturated zone)

Bromine-82, a beta-gamma emitter (Gamma energies: several, ranging from 0.55 to 1.46 MeV) with a half-life of 35.7 hours (powerful injected tracer as $NH_4\,^{82}Br$ for a variety of applications such as seepage studies in dams and reservoirs and for stream flow measurements)

Iodine-131, a beta-gamma emitter (Main gamma energy: 0.36 MeV) with a half-life of 8 days (good gamma tracer for a variety of applications of several days duration, but appears to get retarded in clayey soils)

1.2.4 How are Radioisotope Radiations Detected and Radioisotopes Assayed?

Most of the great advances made in the use of radioisotopes for societal benefit including water resources development and management could not have been possible without parallel development in nuclear electronics (nucleonics) leading to the availability of a host of nuclear detectors. The basic principle of radiation detection is based on the fact that ionizing radiation, while passing through any material; gas, liquid or solid interacts with the orbital electrons of the material. This interaction may cause ionization or excitation of the material resulting in an electrical signal or a light signal. Efficient collection of these signals and converting them into electrical outputs is the basis of radiation detection.

Based on the type of signal generated and electronic processing of the signal, some simply record the total number of interacting particles/gamma photons (e.g.: Geiger Müller counters) and some others provide additional information on the energy distribution of the interacting radiations (e.g.: scintillation and semiconductor spectrometers). In all cases, the detector output is expressed as count rate (Counts per minute/ second or simply cpm/cps). This value is proportional to the level of radioactivity in the sample. Since we live in a naturally radioactive environment, all detectors do show some level of background radioactivity, without any sample. It is important to determine the background levels as we normally deal with low levels of radioactivity in hydrological investigations. The background has to be subtracted from the value obtained from the sample to arrive at the net radioactive level of the sample.

In practice, portable and battery operated sodium iodide scintillation detectors/ spectrometers are used for fieldwork involving gamma-emitting radiotracers like bromine-82. Here the sodium iodide crystal produces a light pulse on interaction with a gamma ray and the light pulse is, in turn, converted to an electrical pulse through a photomultiplier. (See Fig.1.3)

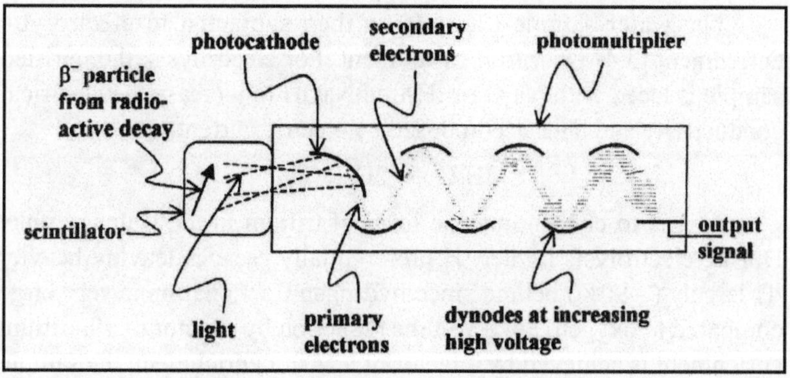

Figure.1.3. Scintillation counter operation

Beta emitting tritium and carbon-14 cannot be assayed in solid scintillation detectors. Liquid scintillation counters are used for this purpose. The radioactive sample is directly introduced into a vial containing a special liquid scintillation cocktail and then the detection procedure is similar to the above.

The reader may refer to Bibliography given at the end of this chapter for detailed information on radiation detection and assay.

1.3 Sampling and Assay of Tritium and Carbon-14 in Natural Water Samples

Tritium Analysis

Natural tritium levels are normally very low for direct measurement even with special low background liquid scintillation detectors. Cosmic ray produced tritium levels in Indian waters are of the order 6-8TU. (1TU or Tritium Unit is defined as one atom of tritium (^3H) in 10^{18} atoms of hydrogen. This works out to be 0.118 Bq/L or 3.19 pCi/L).

Hence a large water sample is normally needed. 500mL samples are normally collected in dry high-density polyethylene or glass bottles and screw capped. Glass bottles, though inconvenient for transport are preferred if a long-term storage of a year or more is envisaged.

The water sample is first purified by distillation to remove impurities, which may later interfere during liquid scintillation counting of tritium. The distillation has to be quantitative (i.e. complete) to avoid any isotope fractionation.

The water sample (500mL) is then subjected to electrolytic enrichment [3,4] prior to measurement. For electrolysis, the purified sample is laced with Na_2O or ^3H-free NaOH to increase the electrical conductivity and later decomposed by electric current.

$$2H_2O \rightarrow 2H_2 + O_2$$

This is to concentrate the level of tritium in the water sample. During electrolysis lighter ^1H preferentially escapes leaving heavier ^3H largely (~ 90%) behind since hydrogen fractionation is very large compared to oxygen. For a volume reduction by a factor of 10, tritium enrichment is achieved by a factor of 9. After enrichment, the sample is again distilled to remove the NaOH or Na_2O added earlier.

The enriched and purified sample is assayed in a liquid scintillation counter. It is normally possible to achieve a detection limit of <1TU with an enrichment factor of about 35 and at a background level of 2 to 3 cpm in a liquid scintillation spectrometer.

Carbon-14 Analysis

As mentioned earlier carbon-14 exists in water as dissolved carbonate or bicarbonate. There are two methods of collection of dissolved inorganic carbon depending on the carbonate level in water.

If the carbonate level is low, the carbonate is extracted from about 50 L of water by acidifying the water and absorbing the evolving CO_2 in an alkaline solution. In the other case, the carbonate is directly precipitated as $BaCO_3$ by addition of NaOH and $BaCl_2$. At the laboratory, CO_2 is evolved from the field carbonate with phosphoric acid and either of the following procedures is used to analyze for carbon-14.

Benzene synthesis: The gas is first converted to acetylene (C_2H_2) and then trimerised to benzene (C_6H_6). The benzene sample is then assayed in a low background liquid scintillation counter [3].

CO_2 absorption: The gas is directly absorbed in a special scintillation cocktail and assayed in low background liquid scintillation counters [5]. The special cocktail consists of a mixture of equal quantities of Carbo-sorb and Permafluor, both from Packard. 1g of carbon can be taken in 24 mL of the absorbing medium. With this method it is possible to achieve a detection limit of about 1 per cent Modern Carbon (pMC). Modern Carbon = ~ 0.2 Bq/gC (per gram of carbon). This means it is possible to date waters with ages of about 35,000 years.

The original ^{14}C standard was NBS Oxalic acid (Ox1) of the year AD1950 that had an activity of 0.238 Bq/gC. For hydrological purposes, ^{14}C standard activity is defined as 95% of that value which works out to be 0.226 Bq/gC.

1.4 Environmental Stable Isotopes

As mentioned earlier, both injected and environmental radioisotopes are of interest in isotope hydrology investigations. But, in the case of stable isotopes, they are of no practical interest as injected tracers, but of immense value as environmental tracers. Let us look at the stable isotopic forms of the water molecule as well as those of the dissolved salts.

Isotopic forms of the water molecule

Hydrogen has 3 isotopes and oxygen also has 3 isotopes in nature. They are

$$^{1}H, ^{2}H, ^{3}H \text{ and } ^{16}O, ^{17}O, ^{18}O$$

(The super script on the left is the mass number of the isotope)

^{1}H is also called protium, ^{2}H as deuterium and ^{3}H as tritium. So very often the symbol ^{2}H is replaced by 'D' and ^{3}H by 'T'. In this book, the reader may frequently come across D in place of ^{2}H and T in place of ^{3}H.

$^{1}H^{1}H^{16}O$, $^{1}H^{2}H^{16}O$ or HDO, $^{1}H_2{}^{18}O$ are the three stable isotopic forms of water molecule of practical interest in hydrology. $^{1}H^{1}H^{17}O$ and other possible isotopic forms are of no practical interest. The natural abundances of the four isotopes in the hydrosphere are as follows:

Isotope	Natural abundance (ppm)
^{1}H	999850
^{2}H	150
^{16}O	997590
^{18}O	2050

Environmental Stable Isotopes of Dissolved Salts – Used in Hydrology

^{13}C in dissolved carbonates and bicarbonates

^{34}S in dissolved sulphates

The natural abundances of these isotopes are as follows:

Isotope	Natural abundance (%)
^{12}C	98.9
^{13}C	1.1
^{32}S	95.0
^{34}S	4.2

The other stable isotopes of sulphur, ^{33}S & ^{36}S, which are less abundant, are not of significance to hydrology.

Stable isotopes are characterized by their mass numbers i.e. Proton number + Neutron number. It is the mass-wise isotope abundance ratio or simply the **isotope ratio (R)** of the sample that is of interest. It is defined as

$$R = \frac{\text{Abundance of rare isotope}}{\text{Abundance of abundant isotope}}$$

Examples are **$[^{2}H]/[^{1}H]$, $[^{18}O/^{16}O]$, $[^{13}C/^{12}C]$, $[^{34}S/^{32}S]$**

Stable isotope compositions in a given sample are normally reported as per mil deviations (δ) from an internationally accepted standard.

For example, the hydrogen isotope composition is given as

$$\delta^2H\ (‰) = \{(R_{sample} / R_{std}) - 1\} \times 1000$$

where R = [$^2H/^1H$] i.e. ratio of abundance of rare isotope to the abundance of most abundant isotope. $\delta^{18}O$, $\delta^{13}C$ and $\delta^{34}S$ are similarly defined.

Hence an Isotope Ratio Mass Spectrometer (IRMS) is the main instrument for the analyses of stable isotope composition of water samples in hydrology.

Standards

The standard for ^{2}**H** and ^{18}**O** measurements in water was a non-existent Standard Mean Ocean Water (SMOW), which in fact, had an average value of all oceans of the world. Later the International Atomic Energy Agency (IAEA), Vienna made available for distribution a standard close to SMOW and called it **V-SMOW**.

The standard for ^{13}**C** measurements in carbonate was PDB (Pee Dee Belemnite), a carbonate found in the Cretaceous Pee Dee formation in North America. After it was exhausted, a limestone standard that was compared with the original PDB by an IAEA panel and now called ***VPDB*** is being used. The difference between the two is negligible.

Similarly for ^{34}**S** measurement, a new standard **V-CDT**, close to the original CDT (Canyon Diablo Troilite), the FeS phase of an iron meteorite collected around Meteor Crater, Arizona in USA, is established

The following is a summary of the information on the standards.

For ^{2}H in water samples: V-SMOW (Vienna-Standard Mean Ocean Water)

$^2H/\,^1H$ in V-SMOW: 155.75 ppm

For ^{18}O in water samples: V-SMOW

$^{18}O/\,^{16}O$ in V-SMOW: 2005.2 ppm

For ^{13}C in carbonates: PDB (Pee Dee Belemnite)
$^{13}C/^{12}C$ in VPDB: 1.12%

For ^{34}S in sulphates: CDT (Canyon Diablo Troilite)
$^{34}S/^{32}S$ in V-CDT: 4.5%

1.4.1 Sampling for Measurement of Stable Isotope Ratios

Sampling for $^{2}\mathbf{H}$ and $^{18}\mathbf{O}$ is relatively simple. Sample size needed is about 50 mL. Clean and dry screw capped bottles, preferably with a secondary 'poly-seal' cap are filled almost to the brim (allowing for possible thermal expansion). Plastic bottles are normally adequate and are better suited for transport. If long-term storage is envisaged, it is better to opt for glass bottles. For snow and ice, the sample may be first collected in an airtight plastic bag, allowed to melt and then poured into bottles as mentioned above. In all cases, the objective is to avoid any evaporation and exchange with atmospheric moisture after sampling and before measurement. It is prudent to have some percentage of duplicate samples to allow for possible damage.

In the case of flowing streams and shallow ponds, a surface sample is normally representative whereas in the case of deeper lakes, groundwater, snow and ice where stratification is possible, it would be necessary to collect samples at different depths.

For $^{13}\mathbf{C}$ analysis, barium carbonate is precipitated from 50 L of water in the field using the same procedure as described above for ^{14}C analysis (see Fig. 1.3).

For $^{34}\mathbf{S}$ analysis, water samples are acidified and then the sulphate is precipitated using $BaCl_2$ solution. In the case of low sulphate samples, the water sample is pre-concentrated on an ion-exchange column before precipitating the sulphate.

1.4.2 Sample Preparation for Mass Spectrometric Measurement

As already mentioned, we need to use gas source isotope ratio mass spectrometers for the measurement of stable isotope levels in all the above-mentioned samples. Hence the samples need to be converted to appropriate gases before introducing into the mass spectrometer.

For $^{2}H/^{1}H$ measurement, two methods are in use. One is to quantitatively reduce the water sample to hydrogen gas by one of the following reactions

$$H_2O + Zn \rightarrow ZnO + H_2 \quad \text{at } 500°C$$
$$H_2O + U \rightarrow UO_2 + H_2 \quad \text{at } 800°C$$

The other is an equilibration method in which the $^2H/^1H$ ratio of the water sample is equilibrated with that of hydrogen gas through isotope exchange. This is achieved by keeping the water sample agitated for a period of time in hydrogen atmosphere.

$$\mathbf{^2H\,^1HO + {}^1H \rightarrow {}^1H_2O + {}^2H\,^1H}$$

It is the equilibrated hydrogen gas that is introduced to the mass spectrometer for analysis.

In the second method, isotope fractionation occurs. Since the international standard will also be subjected to the same equilibration, the δ value will not vary. However, a good control over the temperature of equilibration is necessary as the isotope fractionation is very sensitive to temperature (6‰ per °C)

For $^{18}O/^{16}O$ measurement, similar equilibration method as above is used. Here a few mL of water is equilibrated with 10 to 20 mL of carbon dioxide at about 25°C.

$$\mathbf{H_2{}^{18}O + C^{16}O_2 \rightarrow H_2{}^{16}O + C^{18}O^{16}O}$$

Here again, it is the equilibrated carbon dioxide that is introduced to the mass spectrometer for analysis.

For $^{13}C/^{12}C$ measurement, the carbonate sample obtained from the field is converted to CO_2 with phosphoric acid and the gas is used for mass spectrometric assay.

For $^{34}S/^{32}S$ measurement, the sulphate sample from the field is converted to SO_2 for mass spectrometric measurement.

1.4.3 Mass Spectrometric Analysis of Stable Isotopes

The working principle of a mass spectrometer of our interest is as follows. Please (see Fig. 1.4).

The mass spectrometer consists of an ion source to ionize the gas sample, an electrical field to accelerate the ions, a magnetic field perpendicular to the electric field to bend the ions into circular paths of different radii depending on their masses and finally an ion collector for each ion beam of given mass value. It can be shown that the radius r of any ion beam is proportional to the square root of the ion mass m, i.e. r (:)√m. The collectors produce electrical currents (I) proportional to the intensities of the beams and the ratios of these currents are converted into isotope ratios.

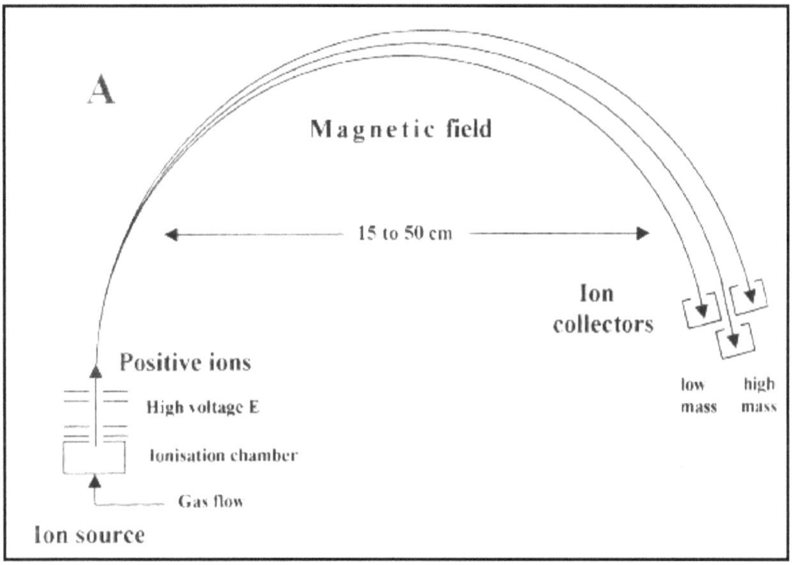

Figure 1.4 Principle of mass spectrometer.

In the case of measurement of $^2H/^1H$, the molecular ion beams of interest are $^2H^1H$ (mass 3) and $^1H^1H$ (mass 2). Complication arises by the formation of $^1H_3^+$ (also of mass 3) by the combination of H_2 gas with a hydrogen ion. Necessary correction is made by a method of extrapolation to zero gas pressure in the ion source where $^1H_3^+$ disappears.

$$\frac{(I_3/I_2)_{sample}}{(I_3/I_2)_{ref}} = \frac{(^2H^1H/^1H_2)_{sample}}{(^2H^1H/^1H_2)_{ref}}$$

Hence

$$\delta^2H = \left[\frac{(^2H/^1H)_{sample}}{(^2H/^1H)_{ref}} - 1\right] \times 10^3 ‰ = \left[\frac{(I_3/I_2)_{sample}}{(I_3/I_2)_{ref}} - 1\right] \times 10^3 ‰$$

In the case of $^{18}O/^{16}O$ measurements in water samples we are concerned with ion beams of $^{12}C^{16}O_2$ (mass 44) and $^{12}C^{18}O^{16}O$ (mass 46) and the ion intensity ratio of (I_{46}/I_{44}). Similar procedure as above is followed to arrive at $\delta^{18}O$.

For $\delta^{13}C$ in carbonates the ion intensity ratio of interest is (I_{45}/I_{44})

In practice, sample and reference gases are separately introduced into the mass spectrometer. The δ value for each sample is usually based on 3 or more sets of measured ratios. Each set consists of 6 reference and 6 sample ratios.

$$\delta = (X_m/R_m)\,10^3\,‰$$

where X_m is the mean difference between sample and reference isotope ratios and R_m is the mean of the reference ratios.

It is not always possible to use an international standard for routine analyses. A secondary laboratory standard that is calibrated against an international standard is normally used. For example, tap water could be a secondary standard for hydrogen and oxygen isotopes in water. Commercially available CO_2 and SO_2 are used similarly for ^{13}C and ^{34}S measurements

It is possible to obtain accuracies better than 1‰ for δ^2H and 0.1‰ for $\delta^{18}O$. These accuracies are adequate for interpretation of hydrological phenomena.

References

1. H. Beherns (1983), "Comparison of radioactive and non-radioactive tracers", Tracer Methods in Isotope Hydrology, IAEA-TECDOC-291, IAEA, Vienna, pp 173-185.

2. S.M. Rao (1984), "Injected radiotracer techniques in hydrology", Proc. Indian Acad. Sci. (Earth Planet. Sci), Vol.93, pp 319-335

3. Environmental isotopes in the hydrological cycle – Principles and applications, (Ed: W.G. Mook), IHP – V, Technical Documents in Hydrology, No.39, Vol.1, UNESCO, Paris 2000, pp 174-177.

4. A.R. Nair, T.B. Joseph, S.M. Rao, (1990) " Measurement of low level tritium in water by electrolytic enrichment and scintillation counting", Bulletin of Radiation Protection, Vol.13, No.1, pp 43-46.

5. A.R. Nair, U.K. Sinha, T.B. Joseph, S.M. Rao, (1995), "Radiocarbon dating up to 37000 years using CO_2 absorption technique", Nucl. Geophys. Vol.9, N0.3, pp 263-268

BIBLIOGRAPHY

1. "Nuclear and Radiochemistry" G. Friedlander, J.W. Kennedy, E.S. Macias, J.W. Miller; Wiley Int. (3rd Edition) 1981
2. "The Atomic Nucleus", R.D. Evans; Tata McGraw Hill (1978)
3. "Nuclear radiation Detection", W.J. Price, (2nd Edition); McGraw Hill, New York, 1964
4. "Radiation Detection and Measurement" G.F. Knoll (3rd Edition); John Wiley and Sons, New York 1999
5. "Stable Isotope Hydrology – Deuterium and Oxygen-18 in the Water Cycle" (Ed. J.R. Gat, R. Gonfiantini), Technical Reports Series No. 210, IAEA, Vienna, 1981
6. "Isotopes in Catchment Hydrology" (Ed. C. Kendall, J.J. McDonnell), Elsevier, 1998

2

BASIS FOR APPLICATIONS OF VARIOUS ISOTOPES IN HYDROLOGY

In the last chapter, we have broadly introduced the concept of tracer hydrology and also the concepts of isotopes as tracers, their characteristics and measurement techniques. In this chapter let us see the basis of application of each group of isotope tracers, injected and environmental.

What are the different groups of isotope tracers used in hydrology and how are they classified?

2.1 Isotope Tracers in Hydrology

Injected

Conservative (radioactive tritiated water $^{1}H^{3}HO$ or HTO)

Non-conservative (radioactive isotopes ^{82}Br as NH_4Br, ^{60}Co as $K_3Co(CN)_6$, ^{131}I as KI *etc.*)

Environmental

Conservative (isotopic species of the water molecule, both stable and radioactive; $^{1}H^{2}H^{16}O$ or HDO, $^{1}H^{1}HO^{18}$, $^{1}H^{3}H^{16}O$ or HTO)

Non-conservative (stable ^{13}C and radioactive ^{14}C in dissolved inorganic carbon and stable ^{34}S in dissolved sulphates)

Conservative tracers faithfully follow water through the hydrological cycle without any significant chemical, geochemical or biogeochemical interactions other than what the water itself is exposed to during phase transitions and at high temperatures (*e.g.* Geothermal waters). So, no specific correction may be necessary while interpreting the conservative tracer data. On the other hand, behaviour of non-conservative tracers need to be interpreted based on their possible interaction with the environment. Halides like bromide may have only negligible interaction whereas some metal complex ions with higher valence state may be influenced by anion exclusion [1] in clayey soils, carbonates by isotopic exchange with aquifer matrix and sulphates by bacterial reduction to sulphides under anaerobic conditions.

2.2 Basis for Applications of Injected Radiotracers

Radioisotope tracers or radiotracers revolutionized tracer technology in various fields such as medicine, agriculture, biotechnology, industry and hydrology. Nuclear medicine is a well-established field all over the world. Radioisotope-labeled pharmaceuticals are used as tracers to study organ functions for disease diagnoses. So is the case with fertilizer uptake studies and investigations on fate of pesticides in agriculture, troubleshooting and process control in industrial plants and a host of investigations in biotechnology.

What makes radioisotope labeled compounds ideal tracers for a variety of applications including hydrology?

Firstly, the behaviour of chemical compounds such as dyes and salts is dictated by their molecular or ionic structure. On the other hand, since isotopy or radioactivity is an elemental property, we have a wide choice of molecular structures into which the desired isotope or radioelement can be incorporated. For example, bromine-82 as ammonium bromide is a good hydrological tracer; as paradibromo-benzene, it is an excellent tracer for many petrochemicals and finally as methyl bromide gas it is a convenient gas tracer. ^{60}Co as chloride is not a good tracer for water as the isotope will be in cationic form, but when incorporated in a highly stable metal complex like potassium cobalticyanide, it will be part of a trivalent anion $[^{60}Co(CN)_6]^{3-}$ which is a very good tracer for water.

Secondly, radioisotopes are most sensitive to detection. Radiotracer detectability is several orders better than that of most ionic tracers. They are often detectable at dilution levels of 10^{12}. Thus, a minute amount of tracer is sufficient to generate the basic dynamic response in a hydrological system and it can be introduced as a practically ideal pulse without any disturbance to the system.

The convenience of using minute quantities of tracer results in concentrations several orders below maximum permissible levels with negligible radiological risk. There is thus no major problem of getting the necessary clearances from competent authorities. It must be pointed out the users of artificially injected radioactive isotopes need to be duly licensed.

Thirdly, if gamma-emitting tracers like bromine-82 are used, *in situ* detection is possible without any need for laboratory support. This is because of the high penetrability of gamma radiation and possibility of detection by battery operated portable sodium iodide scintillation detectors.

Fourthly, some of the radioisotopes are short lived and decay fast enabling repetition of the investigations, if necessary.

Following table puts together the characteristics of some commonly used injected radioisotope tracers.

Isotope	Half-life	Energies# (MeV)	Minimum detectable Concentration+ Bq/mL	Useful chemical composition
^{3}H	12.26a	0.018 (Max)	0.01*	HTO ($^{1}H^{3}HO$)
^{51}Cr	27.8d	0.3 (9%)	0.1	^{51}Cr-EDTA
^{60}Co	5.3a	1.17, 1.33	0.001	$K_3\,^{60}Co(CN)_6$
^{82}Br	36h	0.55 – 1.46	0.001	$NH_4\,^{82}Br$
^{131}I	8d	0.36 (Main)	0.01	$K\,^{131}I$

\# The radiation of interest is gamma except for tritium (^{3}H) where it is β^-,

* Without enrichment and using liquid scintillation counter, + Order of magnitude

From a purely physico-chemical point of view, tritiated water or HTO is an ideal tracer as it is an isotopic species of the water molecule itself. There are, however, other practical considerations, which restrict its use.

- Environmental tritium (natural or of thermonuclear origin) is of large value in age dating of young waters. It is prudent to avoid artificial injection of tritium in such areas where the environmental tritium is of value.

- As a beta emitter, tritium is not detectable *in situ* and needs laboratory support.

2.2.1 Application Approaches

Some applications are usually intended for obtaining qualitative or semi-quantitative information. A good example is to verify interconnection between different water bodies such as seepages in dams contributing to a surface stream or to a groundwater body. Another example is to find the direction of groundwater flow in a given location, particularly in hard rocks including karstic aquifers. It will be seen later that it is often desirable to obtain prior information on environmental isotope composition and major ion levels before undertaking an injected tracer investigation. Let's call them Group I applications [2]

The second group (Group II) of applications pertains to flow measurements in localized hydrological regimes and need some simple mathematical formulations based on tracer balance and mass balance approach. Here the tracer-water system is usually well understood and is not plagued with the problem of "sources and sinks". Examples of this type of applications are stream discharge measurements, estimation of rate of soil moisture movement in the unsaturated zone and determination of groundwater velocities.

2.2.1.1 *Principle of Stream Discharge Measurement with Tracers*

Current meters are most commonly used in conventional stream flow measurement. The cross-section of flow is divided into a number of segments. The current meters are used to measure flow velocities in these segments. If velocity is v_i in the segment i of cross-sectional area a_i then the total stream flow rate Q is given as

$$Q = \Sigma\, v_i\, a_i$$

There are situations in which it may not be easy to use a current meter. Examples are shallow streams, low velocities (<5cm/s) and hilly torrents. Secondly, a fairly good knowledge of the cross-sectional profile is necessary. Thirdly, the usual 10 – 15% accuracy obtainable with current meters may not be adequate. Tracer methods help to overcome these problems in a limited way.

Let us consider a reach of a stream into which there is no further addition or extraction. A tracer of mass M is injected at the upstream end and tracer concentration C (t) is monitored at the downstream end. Tracer balance can now be written as

$$\mathbf{M = \int v\ (\sigma)\ [\int C\ (t)\ dt]\ d\sigma} \tag{2.1}$$

σ represents the cross-sectional element corresponding to velocity v. But, a good miscible tracer disperses in the system and becomes fully identical in hydraulic properties of stream flow. So, at a distance from injection point called the **mixing distance,** the mixing is complete and satisfies the condition (dQ/Q) = (dM/M) at all points in the cross-section. Hence ∫C (t) dt will be constant all over the cross-section. Thus Eq. 2.1 reduces to

$$\mathbf{M = \int v\ d\sigma \int C\ (t)\ dt\ = Q \int C\ (t)\ dt\ \ or}$$
$$\mathbf{Q = M/\int C\ (t)\ dt} \tag{2.2}$$

It is important to note that there is no term for cross-section in Eq. 2.2. Thus, the tracer method is independent of cross-section. Eq. 2.2 can be further simplified as

$$\mathbf{Q{=}M/C_{ave}\ T} \tag{2.3}$$

Where C_{ave} is the average concentration of a number of samples collected over time T during the passage of the tracer at the downstream point.

Eq 2.3 is used when the total mass of tracer is injected instantaneously. This is called "Instantaneous Injection Method".

The alternative is to inject the tracer of concentration C_i at a constant rate q into the stream for a period of time. At the point downstream where the mixing is complete, the stream will have constant concentration C_f for a period, say t_c. (See Fig. 2.1)

Rewriting Eq.2.3, we have

$$\mathbf{Q = q\ C_i\ t_c /\ C_f\ t_c = q\ (C_i\ /C_f)} \tag{2.4}$$

This is called "Constant Rate Injection Method". The advantage here is that one single measurement of concentration during the plateau period t_c at the downstream point is sufficient.

The main limitation of the tracer method is the requirement of adequate "mixing length" before sampling. Several efforts were made to arrive at an empirical formula to estimate the mixing length.

A formula of the form "$L = a\,b\,Q^{1/3}$" where 'b' is the mean width of the stream and 'a' is a coefficient appears to estimate reasonably well [2].

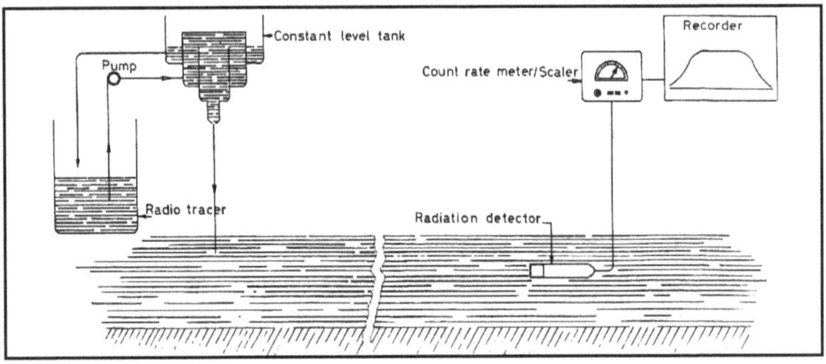

Figure 2.1. Constant Rate Injection Method -Region of constant concentration

Examples of application of the methods are given in Chapter 3.

2.2.1.2 *Estimation of the Rate of Infiltration in the Unsaturated Zone*

The rate of infiltration of rainwater through the unsaturated zone is of interest for setting up a regional water balance. The measurement enables an estimation of the fraction of rainfall that goes to recharge the groundwater reservoir.

There are a number of direct and indirect methods to estimate infiltration rates. Direct methods using lysimeters or infiltrometers are not only cumbersome, but also disturb the natural conditions of flow and may not always be representative. The indirect method of measuring change in the soil moisture profiles needs to be coupled with computation of evapotranspiration with the attendant inaccuracies.

The tritium tracer method [4] and an alternative gamma tracer method [1] offer reasonable direct approaches to estimate rainfall recharge to shallow aquifers.

Before discussing the tracer method, let us see what soil physics says about soil moisture transport.

The Darcy's equation for the movement of soil moisture in the vadose zone can be written [3], as

$$\mathbf{v_f = - K\ [1 + (\partial H/\ \partial m).\ (\partial m/\partial z)]} \qquad (2.5)$$

v_f is the soil moisture Darcy velocity (flow per unit area) in z (downward) direction

H is the soil moisture suction

K is the hydraulic conductivity at soil moisture content 'm'

The hydraulic conductivity K and the Specific water capacity ($\partial m/\partial H$) are both moisture dependent. Difficulties immediately arise for a practical application of the above equation, as we do not know the form of 'K (m)' or of ($\partial H/\partial m$).

If we can assume ($\partial m/\partial z$) = 0 in a portion of the vadoze zone, Eq. 2.5 reduces to

$$\mathbf{v_f = -K} \qquad (2.6)$$

Hence the velocity, which is a moisture dependent property, needs to be determined to evaluate the rate of infiltration.

Tracer approach is the best way to directly determine the velocity. The procedure simply is to tag a soil moisture layer below the root zone and then allow it to move downwards pushed by the infiltrating rainwater. Assuming a 'piston flow' model of soil moisture transport, the tracer displacement 'd' in time 't' is determined. The velocity (d/t) is the true interstitial velocity (v_t) and not the Darcy velocity v_f. The relationship is simply $v_t = v_f/m$.

Moisture flux is then v_t (m) x m. Assuming this to be the result of rainfall during time 't', the percentage rainfall 'r' contributing to recharge is

$$\mathbf{r\% = \frac{100\ v_t\ (m) \times m}{R/t} = \frac{100\ d \times m}{R}} \qquad (2.7)$$

The above involves an implicit assumption that the infiltrating rainwater pushes the moisture ahead of it without overtaking. This 'piston flow model' has been reasonably well verified in many tracer experiments

[2,5]. We will have occasion to elaborate on this in Chapter IV with field examples.

2.2.1.3 Determination of Groundwater Velocities

Darcy's law is based on fluid flow in capillary tubes. It simply states the velocity v_t is proportional to the hydraulic gradient I.

$$\mathbf{v_t = k\ I.}$$

k is a coefficient having the same units as velocity 'v_t', which as mentioned in the earlier section, is called the interstitial velocity.

The flow rate of groundwater through a geometrical cross-sectional area A in the saturated zone is

$$\mathbf{Q = k\rho AI = KAI} \tag{2.8}$$

since the effective area of flow cross-section is the geometrical area multiplied by the porosity 'ρ' of the medium. K is called the coefficient of permeability and has the same units as velocity.

Equation 2.8 is the Darcy equation. The Darcy velocity or the filtration velocity is

$$\mathbf{v_f = KI} \tag{2.9}$$

The relationship between the filtration velocity and the interstitial velocity is

$$\mathbf{v_f = v\rho} \tag{2.10}$$

How do the isotope tracers help in determining the filtration and interstitial velocities?

Radiotracer Point Dilution Technique for Filtration Velocity Measurement

In this method a suitable radiotracer like $NH_4{}^{82}Br$ is injected into an isolated volume U in a borehole and the dilution of tracer concentration due to groundwater flow is monitored with time using an *in situ* scintillation detector. A typical arrangement for the filtration (Darcy) velocity measurement is shown in Fig. 2.2. The tracer in the tracer tank is released remotely.

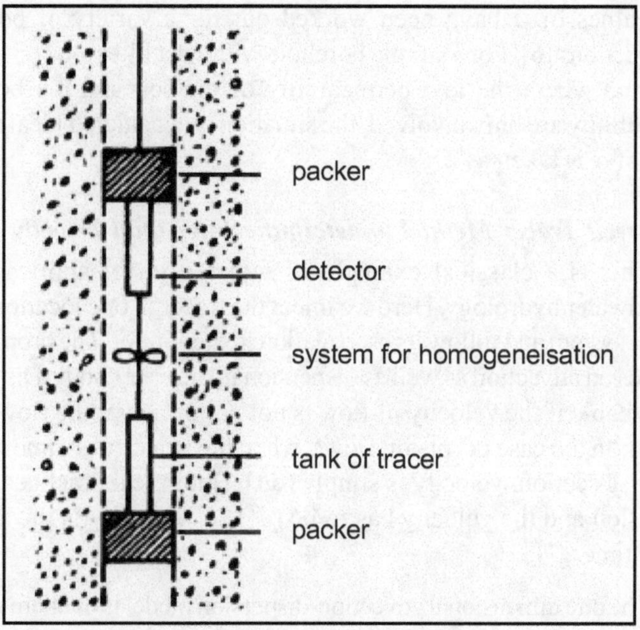

Figure 2.2 A typical arrangement for measurement of filtration (Darcy) velocity by Radiotracer Point Dilution Technique

The rate of dilution is related to the flow rate Q through the isolated portion of the borehole. If the concentration changes from the initial c_0 to c_t in time t, it can be shown [6]

$$\mathbf{Q = (U/t)\ ln\ (c_0 / c_t)} \tag{2.11}$$

The velocity v_b of flow in the borehole is

$$\mathbf{v_b = Q/hD}$$

where D is the diameter of the borehole and h is the height of the isolated section. But, we are interested in the filtration velocity outside the borehole. So, a correction is called for. This correction needs to be based on the hydraulic disturbance to the flow lines due to the presence of the borehole. The correction factor α is defined as $(Q/Q_f) = \alpha$ where Q_f is the filtration flow rate of groundwater through the same cross-section (hD) in the absence of the borehole. So, the final equation for the filtration velocity is

$$\mathbf{v_f = (U/\alpha hDt)\ ln\ (c_0 / c_t)} \tag{2.12}$$

Values of α have been worked out for a variety of borehole constructions [6]. For a simple borehole without a filter tube or a gravel pack and where the low permeability of the soil and the borehole permeability are only involved, the situation is considered ideal and the value of α is taken as 2.

Multi-well Tracer Method to determine Interstitial Velocity

This is a classical example of injected tracer application in groundwater hydrology. Here we inject the tracer in one location in the aquifer system and follow its propagation in the aquifer. The propagation depends on advection as well as dispersion in the water body. Dispersion is inevitable if the velocity of flow is not uniform over the flow cross-section. In the case of 'piston flow', when the velocity is same all over the cross-section, velocity is simply (d/t). But in real practice there is dispersion and the velocity has to be (d/t_{ave}) where t_{ave} is the average transit time.

The one dimensional advection-dispersion model is mathema-tically represented, in its simple form for a conservative tracer, as

$$(\partial c/\partial t) = D_L (\partial^2 c/\partial x^2) - v (\partial c/\partial x) \qquad (2.13)$$

where 'c' is the tracer concentration at distance 'x' from the injection point and at time 't', 'D_L' is the longitudinal dispersion coefficient describing the effects of molecular diffusion and hydrodynamic dispersion and 'v' is the mean interstitial velocity.

There are many solutions to the above dispersion equation based on the boundary conditions. For a uniform pulse injection of an amount 'A' of the tracer, the solution has the form

$$c (x, t) = [A/ (4 \pi D_L t)^{1/2}] \exp [- (x - vt)^2 / 4D_L t] \qquad (2.14)$$

Interpretation of data of tracer dynamics helps us to quantify aquifer parameters such as the mean interstitial velocity and dispersivity 'α'. ($\alpha = D_L /v$)

Dispersivity is an important aquifer parameter to consider while planning pollutant discharge.

The method generally involves pulse injection of the tracer in a borehole and monitoring a set of detection boreholes placed at a reasonably representative distance downstream from the injection hole. (See Fig 2.3)

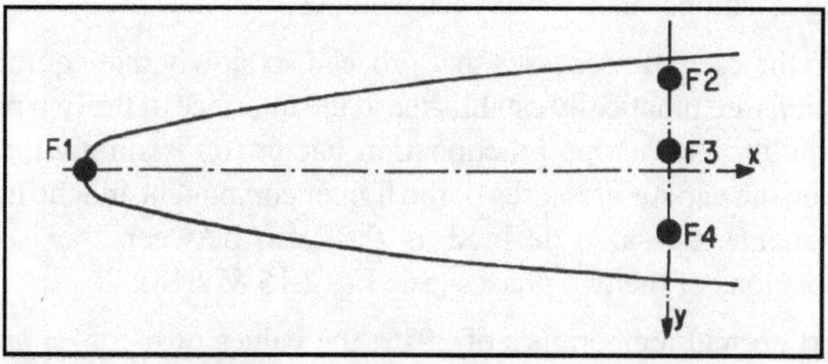

Figure 2.3 An arrangement of injection and detection wells for multi-well method (F1: Injection well, F2, F3, F4: Detection wells downstream, x →: Groundwater flow direction)

Based on concentration versus time '(c – t)' curves recorded at different boreholes, we can obtain the mean velocity of transport, longitudinal dispersion coefficient and dispersivity.

2.3 Basis for the Application of Environmental Isotopes of the Water Molecule

2.3.1 Stable Isotopes; ^{2}H & ^{18}O

As mentioned earlier, one important characteristic of isotopes that makes them useful for many applications in hydrology is the so called isotope effect that causes stable isotopes of the same element having varying rates of participation in physico-chemical reactions.

In chemical reactions, chemical bonds break and form. Introducing a heavier isotope into the reacting molecule can change the rates of breaking and forming of bonds. How much of a change this would cause depends on the ratio of the masses of the heavier and lighter isotopes. The change would obviously be larger for lighter elements compared to heavier elements. This isotope effect is called ***Isotope Fractionation***.

Isotope fractionation is strongly observed in the case of hydrogen and oxygen isotopes in water during phase changes. During condensation and evaporation processes, isotope fractionation occurs due to the higher volatility of lighter $^{1}H_2{}^{16}O$ compared to the heavier isotopic forms, $^{2}H_2{}^{16}O$, $^{1}H_2{}^{18}O$. This results in the enrichment of the heavier molecular species in the liquid phase and their depletion in the vapour phase. Due to their relative mass ratios, the isotope fractionation is larger for hydrogen isotopes than for oxygen isotopes.

In the case of processes that proceed so slowly that equilibrium conditions are practically established at the interface of the two phases, the Equilibrium Isotope Fractionation Factor (α) is simply the ratio between the vapour pressures of the lighter component and the heavier component. It is also defined as the ratio between the isotopic compositions of the two phases (See Eq. 2.15 & 2.16).

At normal temperature of 20°C, the values of α for ^2H and ^{18}O are as follows.

$$^2\alpha_{l/v} = [p\,(^1H_2{}^{16}O)/p'\,(^1H\,^2H\,^{16}O)] = [^2H/^1H]_{liquid}/[^2H/^1H]_{vapour} = 1.084 \tag{2.15}$$

$$^{18}\alpha_{l/v} = [p\,(^1H_2{}^{16}O)/p'\,(^1H_2{}^{18}O)] = [^{18}O/^{16}O]_{liquid}/[^{18}O/^{16}O]_{vapour} = 1.0098 \tag{2.16}$$

In the condensation process, it is the vapour becoming liquid water. So, the relevant fractionation factors are

$$^2\alpha_{v/l} = 1/1.084 = 0.92 \text{ and } ^{18}\alpha_{v/l} = 1/1.0098 = 0.99$$

As is seen above, the isotope effects are in general small, $\alpha \approx 1$. Hence another term *fractionation* ε is defined as

$$\varepsilon = (\alpha - 1)\ x10^3‰,$$

Hence at 20°C

$$^2\varepsilon_{v/l} = -80‰,\ ^{18}\varepsilon_{v/l} = -10‰ \text{ and } ^2\varepsilon_{v/l}/^{18}\varepsilon_{v/l} = 8 \tag{2.17}$$

As mentioned earlier, isotope fractionation is temperature dependent. In an equilibrium process of vapour-to-liquid phase change, $^2\varepsilon_{v/l}$ / $^{18}\varepsilon_{v/l}$ varies from 8.7 at 0°C to about 7.7 at 30°C.

But evaporation, which is a liquid-to-vapour phase change that happens on an ocean surface or any other water body, is a non-equilibrium process as the product is physically isolated from the liquid surface. In such cases the reaction rates and hence fractionation factors are dependent not only on the isotopic masses but also on the vibrational energies of the participating molecules. Such a fractionation is called *kinetic* isotope fractionation.

Reader may consult the bibliography listed at the end of this chapter for a detailed understanding of isotope fractionation.

2.3.1.1 Isotope Fractionation during Precipitation

From a large number of measurements made on precipitation samples from various parts of the world the following global relation has been established [7,8]

$$\delta^2H = 8\delta^{18}O+10 \tag{2.18}$$

This is called Global Meteoric Water Line (GMWL) or simply MWL. This is shown in Fig. 2.4. The line has a slope of 8 and an intercept on the ^2H axis.

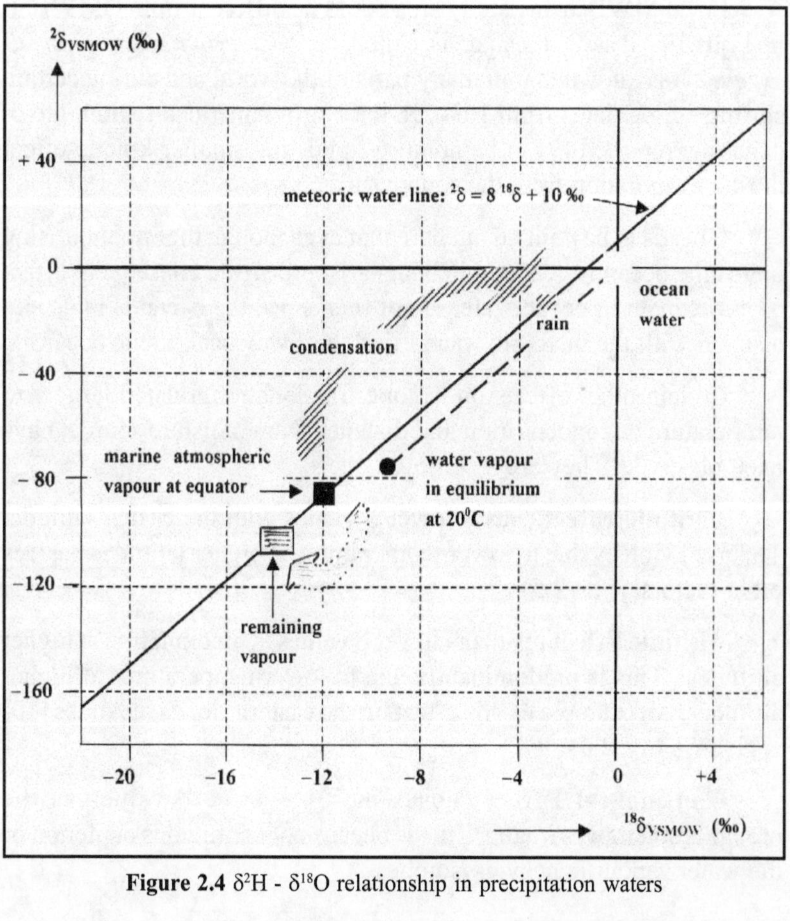

Figure 2.4 δ^2H - $\delta^{18}O$ relationship in precipitation waters

While the slope 8 is explained by our discussion leading to Eq.2.17, the intercept needs to be explained.

Evaporation from the ocean surface is the starting point of precipitation. If this evaporation were also an equilibrium process, then GMWL would have been simply $\delta^2H = 8\ \delta^{18}O$. But, as explained above, it is a non-equilibrium process involving kinetic fractionation. This excess fractionation has a greater effect on lighter hydrogen isotopes than on the oxygen isotopes. This causes the intercept. The general relationship of GMWL is

$$\boldsymbol{\delta^2H = s\delta^{18}O + d} \tag{2.19}$$

Local MWL in any study area could be different from GMWL. If in Eq.II.19 'S' is 8, then 'd' is called the *'deuterium excess' or 'd-excess'*. It is known that in many parts of the world and during certain storms, 'd' deviates from 10‰. In the North American region, the d-excess is reported [9] to be about 6 indicating smaller kinetic effect during evaporation from the source surface.

It needs to be pointed out here that even though the vapour arising out of the ocean is depleted in heavier isotopes, the consequent effect of enrichment in ocean water is not seen since the ocean is huge and also since all the meteoric waters find their way back to the ocean.

Certain other effects on isotope fractionation, related largely to temperature of condensation and distance from moisture source, have been observed. They are as follows.

Latitude Effect causes lower δ values with increasing latitude. This is obviously due to lower temperatures at higher latitudes besides other factors (Fig. 2.5)

Altitude Effect results in lower δ values in precipitation at higher altitudes. This is predominantly due to lower temperatures at higher altitudes. An example of this effect on the eastern slopes of Andes [10] is given in Fig. 2.6.

Continental Effect causes more negative δ values as the precipitation moves inland. This is because of continuous depletion of the water vapour in heavier isotopes

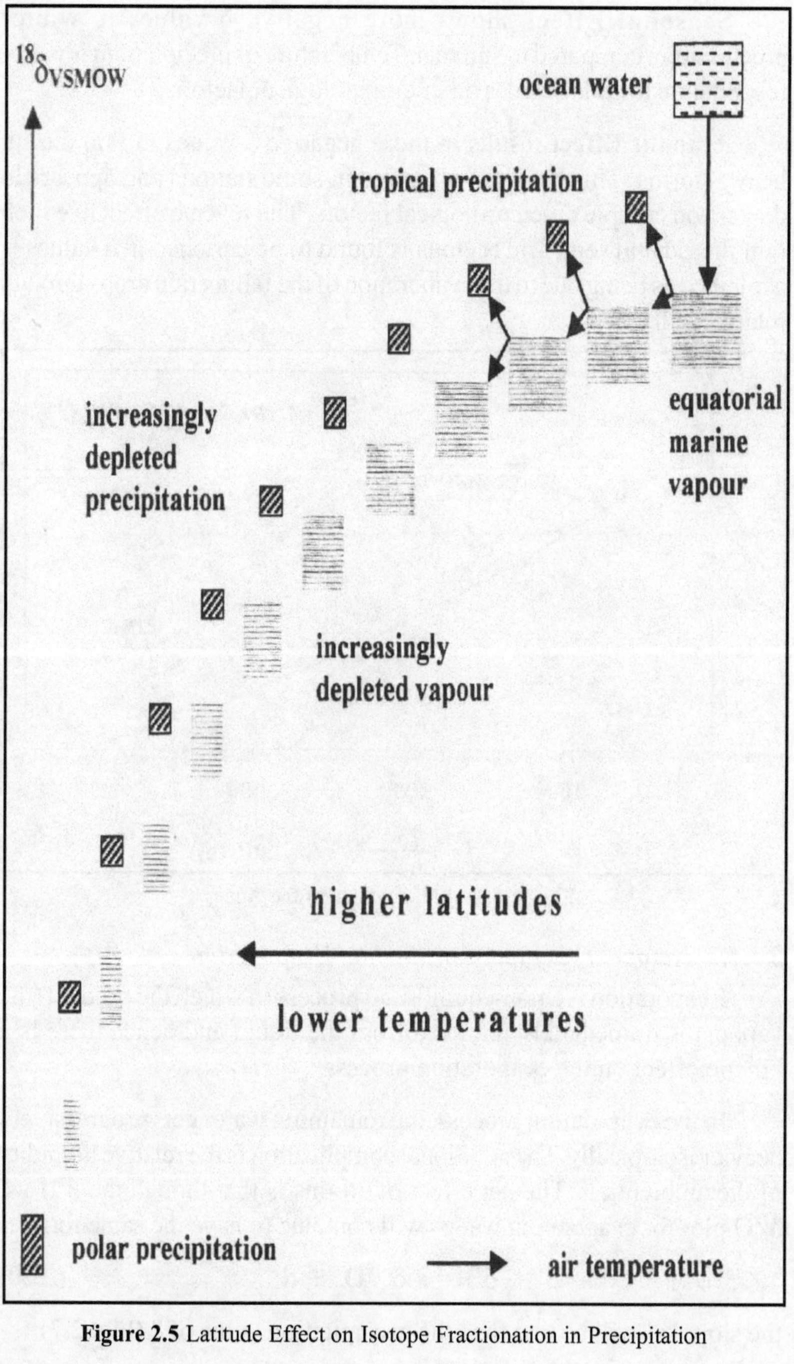

Figure 2.5 Latitude Effect on Isotope Fractionation in Precipitation

Seasonal Effect shows more negative δ values in winter precipitation compared to summer. Temperature of precipitation is partly responsible for this besides other climatological factors.

Amount Effect results in more negative δ values in rain during heavy storms. This has been observed in some stations and appears to depend on complex meteorological factors. The reverse effect in which rain in arid and semi arid regions is found to be enriched in δ values is explained as being due to the evaporation of the falling rain drops through relatively dry air.

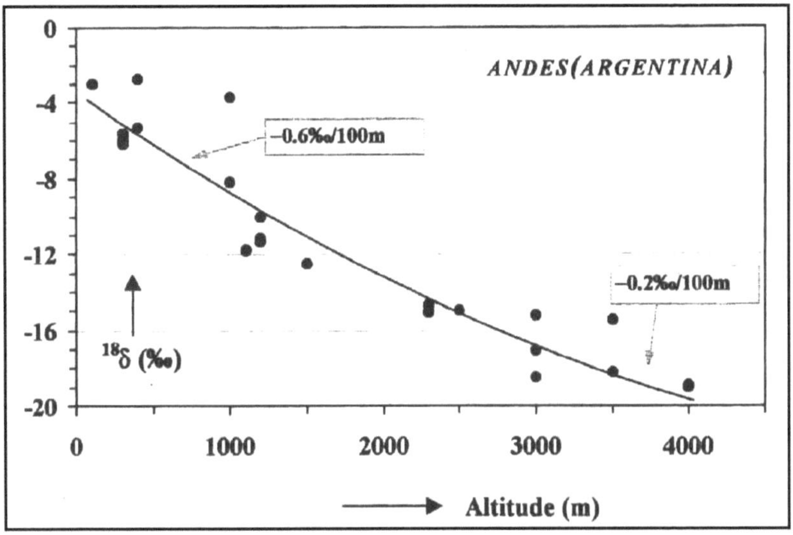

Figure 2.6 Altitude effect on the Andes

2.3.1.2 Isotope Fractionation during Evaporation

Evaporation is a non-equilibrium process in which the product (the vapour) is immediately removed from the liquid and hence there is a kinetic effect on the evaporation process.

In the evaporation process, the remaining water gets progressively heavier isotopically. The additional complication is the relative humidity of the ambient air. The net effect of all this is that though the δ^2H vs. $\delta^{18}O$ plot for evaporating waters will continue to have the same form as

$$\delta^2H = s\,\delta\,^{18}O + d \tag{2.20}$$

the slope 's' will be less than the precipitation value of 8 (Fig. 2.7).

The slope is dependent on the humidity of the ambient air into which the evaporation is taking place. The normal values encountered range from 5 to 7.

The immediate advantage of the δ^2H vs. $\delta^{18}O$ plot is to establish whether a given groundwater body is being recharged directly by rainfall or from an evaporating surface water resource like a lake or a reservoir.

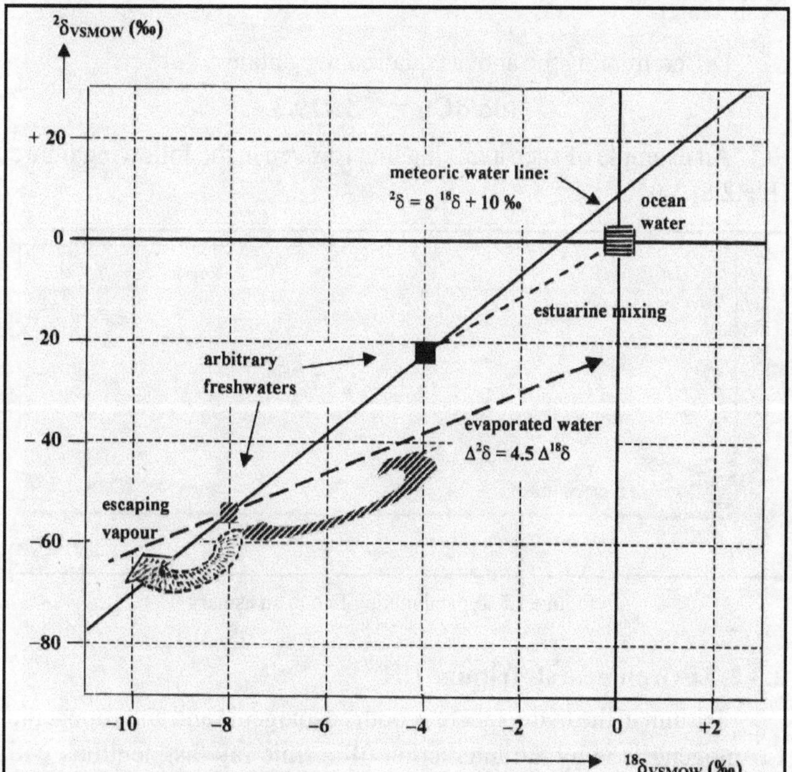

Figure 2.7 Evaporation effect on isotope fractionation

Studying Mixing of Fresh and Saline Waters using ^2H or ^{18}O and Cl^-

The hydrologist is often called upon to establish the source of salinity to a fresh water body. It is possible to demonstrate such connection through isotopic composition by noting whether the stable isotope compositions of the affected waters fall on a mixing line between those of fresh water and a suspected source. This needs to be normally supported by chemistry such as a $\delta^{18}O$ - Cl^- plot. This is possible since isotope composition of the water molecules and chlorinity/salinity are not only conservative, but are also linearly related. For example, in the case of seawater (Chlorinity or Cl ≈ 19.3‰) mixing with fresh water (Cl ≈ 0‰), it can be shown that

$$\delta \approx 1 - (Cl/19.3)\ \delta_f$$

where δ is the isotopic composition of the mixture and δ_f is that of fresh water.

Differentiating the above equation, we obtain

$$(d\delta/dCl) = -\delta_f\ /19.3$$

An example of such a mixing line is given in the following figure. (Fig.2.8)

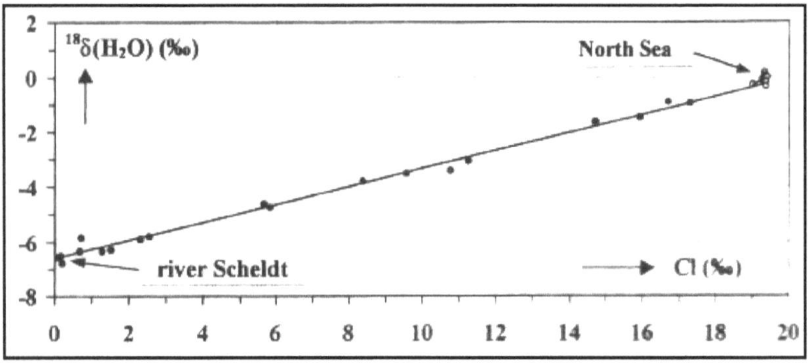

Figure 2.8 Typical mixing line in an estuary

2.3.2 Environmental Tritium (^3H)

Tritium in the hydrosphere is both cosmogenic and anthropogenic. Cosmogenesis is by the interaction of cosmic ray fast neutrons with nitrogen in the upper atmosphere. The reaction is

$$^{14}N + n \rightarrow\ ^{12}C + {}^{3}H$$

(^3H decays to ^3He by emitting a β^- particle with a maximum energy of 0.018 MeV and with a half-life of 12.32 years)

Tritium is then quickly oxidized and enters the hydrological cycle. The tritium concentration in rain due to cosmic ray production is about 5 TU (0.6 Bq/L) in Europe and North America. Values of 6-8 TU are encountered in Indian rainwaters.

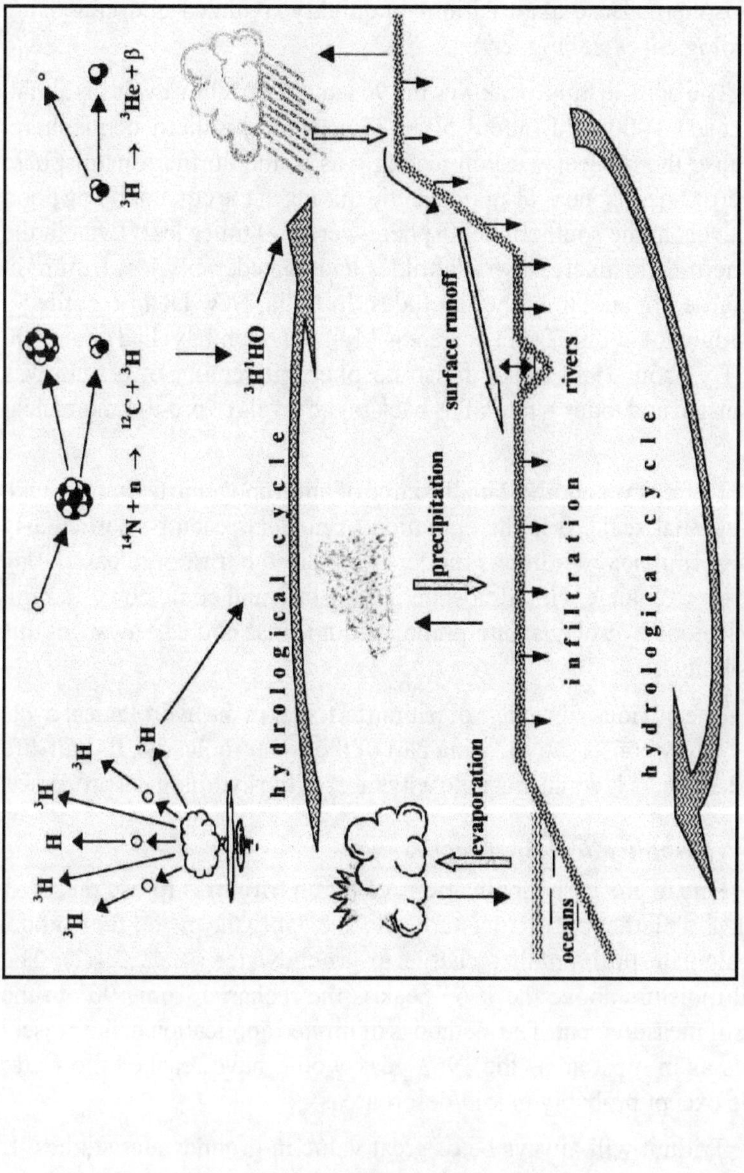

Figure 2.9 Tritium in the hydrological cycle

Anthropogenesis is mainly due to the atmospheric testing of thermonuclear weapons in the fifties and early sixties of the last century before the test ban treaty of 1963. In this, tritium is directly injected from the bombs into the stratosphere and then slowly released to the troposphere. Here also, tritium is quickly oxidized and enters the hydrological cycle (Fig. 2.9).

The bomb tritium peak was in 1963 and the tritium levels in rainfall increased 1000 fold (about 5000 TU) in the northern hemisphere. Because the nuclear weapon testing was in the northern hemisphere and cross-hemispherical stratospheric mixing at the equator being poor, the levels in the southern hemisphere were 100 times less. Even in the northern hemisphere, lower latitudes had considerably less tritium in rainfall compared to higher latitudes. In India, New Delhi recorded a maximum of about 700 TU whereas Mumbai (Bombay) had about 400 TU. By about 1988 or so, the stratospheric inventory of tritium was exhausted and today's rainfall is back to the levels of pre-thermonuclear era.

There is yet another small source of anthropogenic tritium, though on regional scale. It is the operation of nuclear reactors, particularly those using heavy water as moderator. Here the tritium released may quickly exchange with atmospheric moisture and come down in rain. The isotope hydrologist interpreting tritium data should be aware of this possibility.

The obvious advantage of tritium as tracer in the hydrological cycle is it is fully conservative, being part of the water molecule. Its half-life of 12.32 years is commensurate with most of the hydrological processes.

2.3.2.1 Application Approaches

One of the early applications of bomb tritium is to use the 1963 peak as a marker of soil moisture date and apply the piston flow model to calculate the rainfall recharge to groundwater (see 2.2.1.2). The total moisture above the 1963 peak is the recharge from 1963 to the date of measurement. The method is of limited application in the present times as in most areas the 1963 peak would have reached the water table except probably in arid/desert areas.

Tritium will always be of great value in groundwater studies. If detectable tritium is present in a groundwater sample, it means the groundwater body is receiving modern recharge. Though bomb tritium made quantitative dating of groundwater difficult, it is always possible to arrive at some semi quantitative age dating. For example, if the level is < 5 TU, the groundwater surely had a residence time of over 50 years and if the level is > 20 TU, the aquifer was recharged in modern times, after the sixties. It is possible to provide more accurate information through mathematical modeling if we have rainfall tritium data for several years.

Tritium is also a good tool to estimate turnover times of lakes and reservoirs.

In regions of low precipitation, samples from dug wells offer a unique possibility of estimating the upper limit of recharge 'q'. If the aquifer thickness is 'd',

$$\mathbf{q} = \frac{{}^3H_{spl} \cdot d \cdot \rho}{\Sigma\, {}^3H_{in}(t) \cdot e^{-\lambda(now - t)}} \tag{2.21}$$

$[{}^3H_{spl}]$ = Tritium level of well water sample

$[{}^3H_{in}(t)]$ = Tritium input function (precipitation)[11]

ρ = aquifer porosity

2.4 Basis for the Application of Environmental Isotopes of Carbon and Sulphur

2.4.1 *Radioactive Carbon-14 & Dating of Old Groundwater*

Like tritium, ^{14}C is also produced in the upper atmosphere by the interaction of cosmic ray thermal (slow) neurons with nitrogen following the reaction,

$$^{14}N + n \rightarrow {}^{14}C + p$$

(^{14}C decays to ^{14}N emitting a β^- particle with a maximum energy of 0.156 MeV and a half-life of 5730 years.)

^{14}C so formed is quickly oxidized to CO and then to CO_2 and mixes with atmospheric CO_2. In land regions CO_2 is assimilated by plants (See Fig. 2.10) and hence all living organisms, vegetable and animal, contain ^{14}C at the same concentration as atmospheric CO_2 i.e. $[^{14}C/\,^{12}C] = 10^{-12}$ or 0.25 Bq/gC. (disintegrations per second per gram of carbon).

In the literature, we may find several ways of reporting ^{14}C values in groundwater samples. In this book, the chosen way is pMC (percent modern carbon) with atmospheric level of 0.25Bq/gC as 100 pMC.

From partial pressure considerations, the CO_2 dissolved in infiltrating rainwaters is predominantly soil CO_2 from decaying plant remains and root respiration. Fossil carbonate, which is ^{14}C–free is generally part of terrestrial soils. The following reaction results in the formation of bicarbonate.

$$\mathbf{CO_2 + CaCO_3 \rightarrow Ca^{2+} + 2HCO_3^{\,-}}$$

The bicarbonate can have only 50% of ^{14}C from dissolved CO_2. This is the dissolved inorganic carbon (DIC) that is collected for ^{14}C analysis. (See 1.3)

Isotopic exchange with dissolved $^{14}CO_2$ pushes up the initial ^{14}C value of the bicarbonate to a higher value of 65 – 85% of the ^{14}C value of atmospheric CO_2. In the absence of data to quantify isotopic exchange, the initial ^{14}C value is taken as 85% of that of atmospheric CO_2.

The radioactivity of ^{14}C in dissolved bicarbonate and carbonate starts decaying with a half-life of 5730 years and provides a good tool for age dating of groundwater [12].

Recalling the radioactive decay law (Eq. 1.1), we have

$$\mathbf{A_t = A_0\ e^{-\lambda t}}$$

where initial activity A_0 reduces to A_t in time t and λ is the decay constant.

Recharge area geology	A_0 (pMC)	Reservoir Correction (Years)
Crystalline	90 to 100	-1000 to 0
Loss covered	85	-1300
Uncovered karst, dunes	55 to 65	-5000 to -3500

For carbon-14, λ = (ln2/5730). Hence the age t (in years) is given by

$$\mathbf{t = -8267\ln(A_t/A_0)} \qquad (2.22)$$

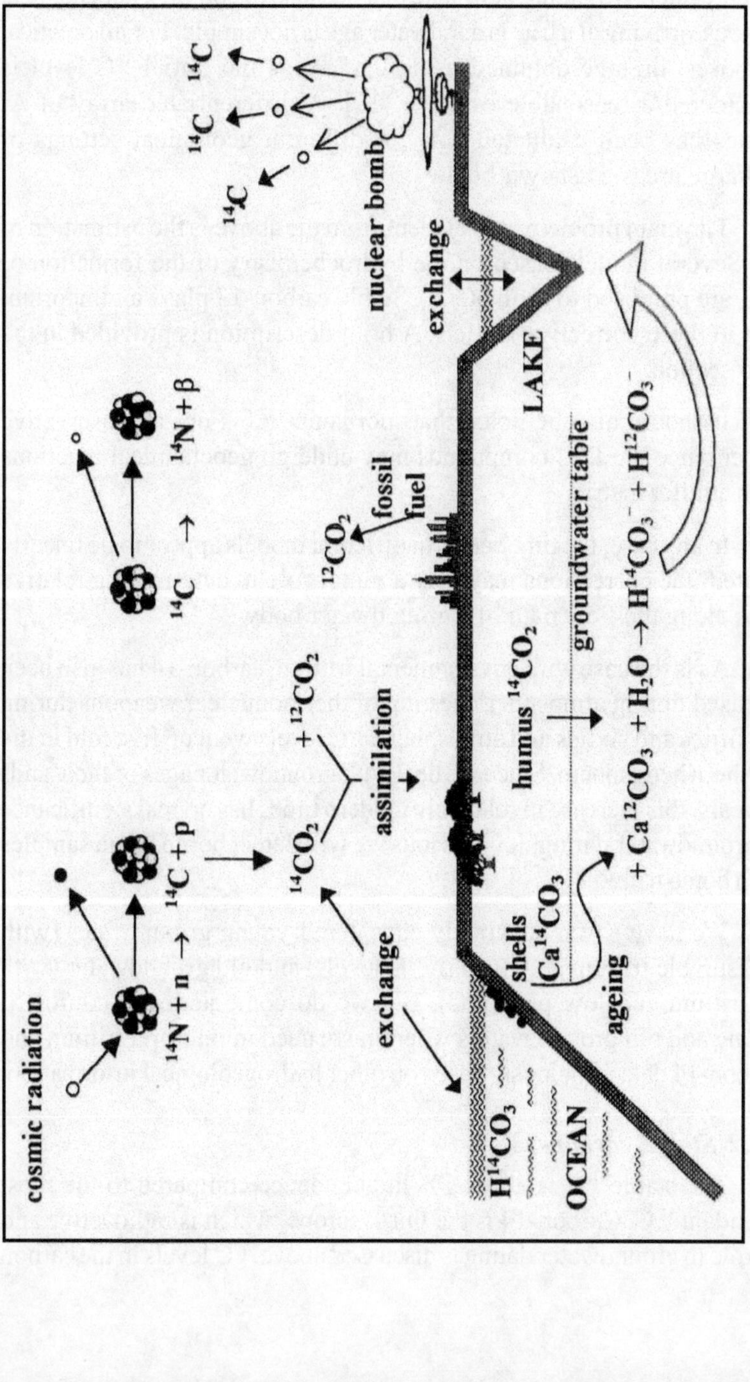

Figure 2.10. Carbon-14 in the hydrological cycle

This is generally referred to as the 'uncorrected age' of groundwater. We should here state that because of complex geochemical reasons, arriving at a true groundwater age is not simple. For all practical purposes, the age obtained using 0.85 A_0 as the initial ^{14}C level is considered a reasonable estimate [13]. A differentiated array of A_0 values has been evaluated [14] for different geological settings of recharge areas as shown below.

The main problem as is evident from the above is the estimation of A_0. Several models based on the hydrochemistry of the formation of DIC are proposed to estimate A_0. Stable carbon-13 plays an important part in these corrective models. A brief description is provided in the next section.

It should also be noted that inorganic ^{14}C is not a conservative tracer since the DIC compounds may undergo geochemical reactions with aquifer matrix.

In any case, the time scales in different models appear to be linearly related, the corrections may play a minor role in determining relative ages along the flow path of a groundwater body.

As is the case with environmental tritium, carbon-14 has also been released during atmospheric testing of thermonuclear weapons during the fifties and sixties and atmospheric ^{14}C levels went up five fold in the northern hemisphere. Since we deal with groundwater ages of thousands of years, this increase in relatively modern times has no real significance in groundwater dating. It is obvious we will detect bomb ^{14}C in samples with bomb tritium.

We have so far separately considered young groundwater (with measurable tritium and modern ^{14}C levels) and old groundwater (with no tritium and low pMC ^{14}C). But we do come across mixtures of young and old ground waters wherein we need to interpret tritium and carbon-14 data judiciously based on other hydrogeological information.

2.4.2 Stable carbon-13

The stable ^{13}C is about 1% in abundance compared to the most abundant ^{12}C. Carbon-14 is the third isotope, which is radioactive and its role in groundwater dating is discussed above. ^{13}C levels in the carbon components like soil CO_2, the soil carbonate and lime in the unsaturated zone are needed to quantify the water-rock interactions and correct ^{14}C ages of groundwater. One commonly used correction procedure to estimate initial carbon-14 level is as follows [15]. This is based on the concept of isotopic mixing (see 2.3.1.3) with ^{13}C contents of soil CO_2 and solid carbonate as the end members.

$$A_0 = \frac{(\delta\ ^{13}C_{DIC} - \delta\ ^{13}C_{lime})}{(\delta\ ^{13}C_{gas} - \delta\ ^{13}C_{lime})} (A_g - A_{lime}) + A_{lime} \qquad (2.23)$$

DIC refers to dissolved inorganic carbon, 'lime' to solid carbonate, gas to soil CO_2. A_g and A_{lime} refer to ^{14}C levels in soil CO_2 and solid carbonate respectively.

Often, this model is further simplified with $\delta\ ^{13}C_{gas}$ = -25‰, a value generally found dicotyledonous plants, $\delta^{13}C_{lime}$ = 0 and A_{lime} = 0. Equation 2.23 then reduces to

$$A_0 = (\delta^{13}C_{DIC}) \ / \ -25 \qquad (2.24)$$

This is generally called the 'Pearson model'

The $\delta^{13}C_{DIC}$ of ground waters can be used to differentiate the recharge zones, say sandstone aquifer ($\delta^{13}C_{DIC}$ = – 15 to –17‰) from a limestone aquifer ($\delta^{13}C_{DIC}$ = -12 to –14‰)

2.4.3 Stable Sulphur-34

Sulphur has four stable isotopes: ^{32}S (95.02%), ^{33}S (0.75%), ^{34}S (4.21%) and ^{36}S (0.02%). The $^{34}S/^{32}S$ ratio expressed as $\delta^{34}S$ with respect to CDT (see 1.4) is of interest in isotope hydrology. There are three main reservoirs of sulphur; evaporite sulphates, marine sulphates and sedimentary sulphides. Hydrological interest is confined to evaporites and marine sulphates. Whereas $\delta^{34}S$ values of evaporites cover a wide range, the value for marine sulphate remained constant at +20‰ for several millions of years.

Groundwater contains dissolved sulphate ions from a few mg/L in shallow aquifers to a hundred times more in fossil brines. The sulphate sources are mainly atmospheric precipitation, mineral/rock dissolution ($\delta^{34}S$ = +5 to 40‰) and infiltration from surface water resources

($\delta^{34}S$ = -10 to +30‰). Atmospheric contribution includes seawater spray ($\delta^{34}S$ = +20‰) in coastal areas and H_2SO_4 in rain due to the burning of fossil fuels ($\delta^{34}S$ =-5 to +10‰). Sulphur isotope fractionation is caused by several geochemical and biochemical reactions (See Fig. 2.11). Most effective among them is the bacterial reduction of sulphates to sulphides under anaerobic conditions in deep groundwater. This results in substantial enrichment of ^{34}S. The fractionation factor ε (SO_4/H_2S) is about 30‰.

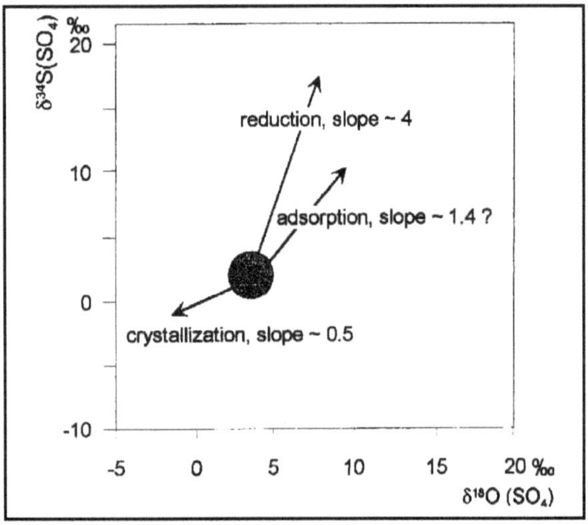

Figure 2.11 Trends in fractionation of ^{34}S and ^{18}O in sulphates [16]

Equally important is the isotope fractionation of oxygen isotopes in sulphates i.e. $\delta^{18}O(SO_4)$. The exchange of oxygen between SO_4^{2-} and water being extremely slow at low temperatures of groundwater, the isotope composition of oxygen in sulphate reflects either mixing or microbial reduction to sulphides.

Since bacterial reduction and consequent enrichment in the isotopic composition being extremely slow processes, any substantial enrichment of ^{34}S is related to a long residence time of groundwater. Presence of H_2S in water and low pMC of ^{14}C are the other confirmatory tests for long residence time.

2.4.4 Other Isotopes of Interest

In this chapter, we tried to understand the characteristics of some isotopes and the basis of their applicability to understand the dynamics of hydrological systems. Most of the field applications are generally restricted to these isotopes

Literature has references to rare applications of environmental isotopes like ^{3}He, ^{36}Cl, ^{37}Cl, ^{39}Ar, ^{85}Kr and ^{81}Kr. As of now, there is some promise for two of them; ^{3}He and ^{36}Cl.

Helium-3

For age dating of young groundwater using bomb tritium, we need to know the tritium input function. We also know that tritium decays to ^{3}He by beta emission. If we can measure tritiogenic ^{3}He concentration in water by suitable mass spectrometry, we can assess groundwater age without the knowledge of the input function. From the decay equation (Eq.1.2), we can show the age 't' in years as

Attention needs to be paid to contribution from atmospheric helium and terrigenic production.

$$t = \frac{1}{\lambda} l_n \left[\frac{^{3}H_e}{^{3}H} + 1 \right] = 17.8\, l_n \left[\frac{^{3}H_e}{^{3}H} + 1 \right] \qquad (2.25)$$

Chlorine-36

Chloride ion is a very attractive tracer chemically as it is practically conservative in all hydrological environments. Unfortunately there is no suitable isotope of chlorine for convenient application in isotope hydrology. Radioactive ^{36}Cl is produced in the upper atmosphere by reaction of protons with argon, but in extremely small quantities (about 26 atoms/m^2/s). It is also produced near the earth's surface by the interaction of cosmic ray neutrons with the most abundant (75.5%) ^{35}Cl and also to some extent by terrestrial neurons in deeper layers. The latter will depend on the concentration of uranium and thorium in the aquifer matrix. Adding to all this is the contribution from thermonuclear weapon tests. The initial $^{36}Cl/Cl$ ratio is of the order of 5 to $30x10^{-15}$. These small concentrations are now amenable to measurement due to recent introduction of Accelerator Mass Spectrometry (AMS) techniques. The measurements are, however, very expensive.

^{36}Cl decays to ^{36}Ar by β^- emission (E_{max} ~0.7 MeV). It has a half-life of 300,000 years. The conservative nature of the chloride ion and the long half-life of ^{36}Cl have made the isotope particularly attractive for age dating of extremely old waters. The interpretation of data is difficult in practice because of the different production processes of the isotope and dilution of the tracer by the dissolution of chloride.

There are only a few examples of ^{36}Cl application. One is the confirmation of the hydraulically estimated groundwater ages of up to a million years in the Great Artesian Basin in Australia [17]. Another is the age dating of groundwater in the Milky River aquifer in Canada. It is here that the difficulties in interpretation of data were fully recognized [18].

References

1. S.M. Rao, S.K. Jain (1985), "Tracer Transport Modelling for Unsaturated Zone – Effect of Anion Exclusion and Immobile Water", Isotopenpraxis, 21, 12, 8 pp 433 – 438
2. S.M. Rao (1984), "Injected radiotracer techniques in hydrology", Proc. Indian Acad.Sci (Earth Planet. Sci), Vol.93, pp 319-335
3. E.C. Childs (1967), Soil Moisture Theory, Advances in Hydroscience, Vol.4, Academic Press, p.89
4. U. Zimmermann, K.O. Munnich, N. Rother, W. Krutz, K. Schubach, O. Siegel (1966), "Tracers determine soil moisture movement and evapotranspiration" Science, 152, pp346
5. R. Rangarajan, R.N. Athavale (2000), "Annual replenishable groundwater potential of India – an estimate based on injected tritium studies" Journal of Hydrology, 234, pp 38-53
6. E. Halevy, H. Moser, O. Zellhofer, A. Zuber, (1967), Borehole dilution technique – A critical review, Isotopes in Hydrology, International Atomic Energy Agency, Vienna
7. H. Craig (1961), Isotopic variations in meteoric waters, Science, 133, pp1702 – 1703
8. W. Dansgaard (1964), Stable isotopes in precipitation, Tellus, 16, pp43 - 438
9. Y. Yurtsever, J.R. Gat (1981), Chapter 6, Atmospheric Waters In: J.R. Gat and R. Gonfiantini (eds.), Stable Isotope Hydrology, Deuterium and Oxygen-18 in the Water Cycle, IAEA Technical Reports Series No.210, Vienna
10. J.C. Vogel, J.C. Lerman, W.G. Mook (1975), Natural isotopes in surface and groundwater from Argentina, Hydrol. Sciences Bull, XX (2): 203-221

11. IAEA, Statistical Treatment of Data on Environmental Isotopes in Precipitation, IAEA Vienna, 1992, 781 pp (STI/DOC/10/331)

12. K.O. Munnich (1968), Isotopen-Datierung von Grundwasser, Naturwiss, 55, pp158-163

13. J.C. Vogel, D. Ehhalt (1963), The use of carbon isotopes in groundwater studies, In: Radioisotopes in Hydrology, International Atomic Energy Agency, Vienna, pp383-395

14. M.A. Geyh (1972), Basic studies in hydrology and ^{14}C and ^{3}H measurements, In: Proc. XXIV Int. Geol. Congr, 11, pp227-234, Montreal

15. F.J. Pearson Jr, B.B. Hanshaw, (1970), Sources of dissolved carbonate in water and their effects on carbon-14 dating, Isotope Hydrology 1970, Proc. Symp. IAEA, Vienna, pp 271

16. H.R. Krause (1987), Relationships between the sulphur and Oxygen isotope composition of dissolved sulphate, In : Studies on Sulphur Isotope Variations in Nature, Paval Proc. Series IAEA, Vienna, pp 31-47.

17. A.L. Herczeg, A.J. Love, L. Sampson, R.G. Cresswell, L.K. Fifield (1999), "Flow velocities estimated from Cl-36 in South-West Great Artesian basin, Australia", Int. Symp. on Isotope Techniques in Water Resources Development and Management, IAEA, Vienna SM-361-34p

18. F.M. Phillips, H.W. Bentley, S.N. Davis, D.Elmore, G.B. Swanick (1986), Chlorine-36 dating of very old groundwater, Milk River aquifer, Alberta, Canada, Water Resourc. Res., 22 (13), pp2003-2016

Bibliography

1. Guidebook on Nuclear Techniques in Hydrology, (1983), Tech. Rep. Series No.91, IAEA, Vienna

2. Environmental isotopes in the hydrological cycle, Principles and applications (Ed. W.G Mook), Introduction, IHP-V, Technical Documents in Hydrology, No.39, Vol.I, Unesco, Paris, 2000

3. Use of Artificial Tracers in Hydrology, Proc. Adv. Group Meeting, Vienna, 1990, IAEA, Vienna (TECDOC-601)

4. Isotope Tracers in Catchment Hydrology (Eds. Carol Kendall, Jeffery J. McDonnell), Elsevier, 1998

3

ISOTOPE APPLICATIONS IN SURFACE WATER HYDROLOGY

The range of isotope applications in surface water hydrology is not as wide as in groundwater hydrology as the surface waters are more easily amenable to conventional hydrological methods. Following are the areas in which isotopes have a significant role in either improving the accuracy of measurements or in providing additional information, which is otherwise not easily available.

- Stream Flow Measurements (^{82}Br, ^{3}H)
- Stream component separation in river flow (^{2}H, ^{18}O)
- Water Balance & Dynamics in river catchments and lakes (^{2}H, ^{3}H, ^{18}O)
- Canal Seepage Studies (^{82}Br, ^{3}H, ^{2}H, ^{18}O)
- Dam Seepage Studies (^{82}Br, ^{3}H, ^{2}H, ^{18}O)
- Surface-Groundwater interrelation (^{2}H, ^{3}H, ^{18}O)

The isotopes normally employed for each application are given in parentheses.

3.1 Stream Flow Measurement

In 2.2.1, we discussed in some detail the principle of stream flow measurement using injected tracers. Let us recapitulate the salient points.

In the tracer method, a quantity 'M' of tracer is injected instantaneously at a point upstream and its concentration distribution with time (C-t profile) is monitored at a point downstream where the tracer is fully mixed with the water body. From tracer balance and mass balance considerations, we showed that the flow rate Q is given by

$$Q = M / \int C(t)\, dt$$

If, on the other hand, the tracer of concentration C_i is injected continuously at a rate 'q', we achieve a uniform concentration C_f across the flow cross-section for a period of time at the monitoring point downstream where the mixing is complete. Then we have

$$Q\, C_f = q\, C_i \text{ or } Q = q\, (C_i / C_f)$$

3.1.1 Experimental Technique

The tracers normally used for flow measurements are gamma-emitting ^{82}Br as NH_4Br and beta emitting tritium (^3H) as tritiated water ($^1H^3HO$).

Instantaneous injection is relatively a simple procedure. Since the injected tracer will have a large quantity of radioactivity and the tracer concentrations at the sampling station will be very small, the experimenter will have to observe precautions to avoid any cross-contamination which might jeopardize the results. This is besides taking precautions from the point of view of radiological safety.

In addition to the above, constant rate injection needs a special gadget to ensure an accurate injection rate all through. This is important as it has a direct bearing on the final accuracy of flow measurement.

For instantaneous injection, the tracer (a few mL) received from the supplier is diluted into a weighed amount (a litre or so). An aliquot is taken and is successively diluted to levels expected at the sampling station and concentration determined using a calibrated NaI scintillation counter. This concentration multiplied by the total dilution factor leads to the total injection quantity M of tracer. Care is taken to ensure that the dilution factor is obtained with high accuracy, preferably by weighing the quantities at successive dilutions.

In the case of continuous injection, in addition to following the same procedure as above to determine the concentration of C_i the injected tracer, the rate of injection q should also be known with high accuracy.

The advantage of gamma emitting ^{82}Br is the possibility of *'in situ'* measurement by simply immersing *a well calibrated* NaI scintillation probe into the stream and connecting it to a battery operated field rate meter/counter. The C-t curve can be easily prepared in the field itself for the instantaneous injection procedure.

The disadvantage of the high-energy gamma emitting ^{82}Br is the heavy protective lead pot in which it has to be transported to the field and the remote handling gadgets needed for the preparation and injection of the tracer. In short, the radiological safety procedures are more elaborate.

The beta emitting tritium tracer is easily transported to the field and injection procedures are also relatively simpler. *In situ* detection is, however, not possible with tritium. Samples have to be taken during the passage of the tracer for later measurement using a liquid scintillation counter in a laboratory. Problems do exist to identify the time of arrival of the tracer and also the time of its departure. It is hence desirable to start sampling well before the expected time of arrival and continue sampling till well after the expected time of departure of the tracer.

One major disadvantage with tritium is it could irrevocably alter environmental tritium levels in the regional groundwater and may deprive the hydrogeologist of one possibly important isotope tool.

3.1.2 Field Examples

The main limitation of the tracer method is the "mixing length". Several efforts [1] were made to arrive at an empirical formula to estimate the mixing length. The formula "$L = a\,b\,Q^{1/3}$" where 'b' is the mean width of the stream and 'a' is a coefficient appears to estimate reasonably well. The following table [1] illustrates this.

A value of a = 10 may be considered adequate for estimating the mixing length in a non-turbulent stream and higher values may be needed for turbulent streams. This could be due to insufficient residence time in a given reach for adequate lateral dispersion in fast flowing (turbulent) rivers. This may not be true in all cases as the mixing length is a complex function of hydraulic roughness, channel geometry, turbulence etc.

Stream	Flow Rate (m^3 /s)	Average Stream Width (m)	Observed Mixing Length (km)	Mixing Coefficient a	Remarks
Tapi, (Surat)	1250	510	43	8	Sheet Flow
Tapi, (Surat)	500	510	27	7	Sheet Flow
Ganga Canal, (Roorkee)	150	70	3	8	Sheet Flow
Tons, (Dehradun)	170	45	3.3	13	Turbulent
Beas, (Manali)	6	20	1	28	Turbulent
Beas, (Manali)	20	20	3.5	65	Very Turbulent

*All streams are rivers except Ganga Canal

It is also clear from the above table that the mixing distance is inconveniently long for very turbulent flows of over 50m^3 /s and for flow rates above 200m^3/s for non-turbulent or moderately turbulent flows. The tracer method is particularly useful to measurements in hilly torrents and also when high accuracies of the order of 1 - 2% are needed since conventional current meters hardly achieve accuracies better than 10 - 15%.

There are other examples of application of tritium tracer for flow measurements in large rivers [2]. In large rivers, flow measurement using conventional method is particularly difficult during high flood periods. The following table lists some measurements in large rivers using tritium.

The objective of stream flow measurements using radiotracers is to build accurate Gauge – Discharge rating curves. One possible application of high accuracy stream flow measurement is to measure seepage losses from unlined canals by making the measurements in successive reaches.

River	Flow Rate(s) m^3/s	Activity GBq
Tana (Kenya)	106, 270, 550	320 - 1600
Rufiji (Tanzania)	850, 1025, 2000	280 - 480
Zimbabwe rivers		
Inyagui	57, 72, 143	
Mazoe	126	120 - 960
Umniati	230	
Umfuli	490, 510	

The overall experience in using radioactive tracers for stream flow measurements, particularly of large flows is severely limited. In addition to the general lack of information to the practicing hydrologists, there is also the need to obtain the necessary clearances from competent authorities to use radioactive tracers in the public domain.

3.2 Stream Component Separation in Rivers

Though rivers form only 0.004% of the fresh water resources on the planet, their easy accessibility and exploitability rendered them very attractive and most early human settlements were along riverbanks. For optimum utilization of this resource, we need to have full understanding of the hydrology of the river system. River flow consists of one or more of the following components.

- Base flow consisting mainly of influent groundwater contributing to river flow
- Current precipitation run-off
- Snowmelt water

At any given time two of the three components predominate. During the rainy season, it is a mixture of run-off and the base flow and during the dry summer months it would be a mixture of snowmelt and the base flow. Environmental isotopes provide an elegant approach to quantify the different components in river flow.

In the case of a two-component system, if f_1 and f_2 are the fractions of the two components, then the mass balance of the flow is

$$f_1 + f_2 = 1 \qquad (3.1)$$

The isotope balance is

$$f_1\delta^{18}O_1 + f_2\,\delta^{18}O_2 = \delta^{18}O_{mix} \qquad (3.2)$$

This equation can be written for ^2H also. For this type of investigation, measurement of either isotope is sufficient.

From the above equations, we can show that

$$f_1 = (\delta^{18}O_2 - \delta^{18}O_{mix}) / (\delta^{18}O_2 - \delta^{18}O_1) \qquad (3.3)$$

$$f_2 = 1 - f_1$$

3.2.1 Case Studies

Rhine

The $\delta^{18}O$ value of the river Rhine at a Dutch station drops from -9.5‰ in winter to about –10.5‰ in summer [3]. If one assumes that the Rhine water at the station represents a mixture of alpine melt water and run-off from the lower part of the basin, the fraction of each component could be estimated from the above equations. The alpine component works out to be 50% in summer and dropping to 20% in winter.

Danube

It is the second largest river in Europe with a catchment area of over 800,000 km^2 and a maximum flow rate of 6500 m^3/s. Fig. 3.1 shows the $\delta^{18}O$ levels in Danube at Ulm and at Vienna and also those of its important tributary Inn at Schärding. At Ulm, the river has a mean $\delta^{18}O$ value of -10‰ whereas at Vienna the value is depleted to about –11.5‰. This is the effect of the Alpine tributaries, most important of them being the River Inn. The long-term $\delta^{18}O$ value of Inn is -12.9‰. Since melting at high elevations occurs in late spring and summer, Inn has minimum $\delta^{18}O$ (-13.8‰) in June and maximum (-12.3‰) in December. This is well reflected in the values in Danube at Vienna.

The mean $\delta^{18}O$ levels in Danube and those of precipitation in Vienna were used [4] to separate the flow into 3 components:

1. Base flow ($\delta^{18}O$ = -10.8 to -11‰)
2. Direct precipitation (for Alpine precipitation at 1500 m altitude, an altitude effect of –0.2‰ per 100 m is assumed, varies from -16‰ in winter to -10‰ in summer)

3. Melt water contribution mainly from April to September ($\delta^{18}O$ = -16‰)

It is shown that the base flow accounts for about 90% of the total flow in winter and it falls to 50% in summer.

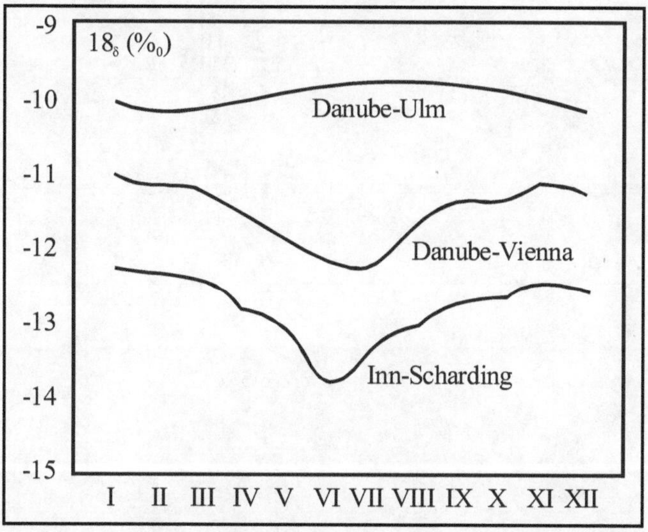

Figure 3.1 Mean monthly $\delta^{18}O$ values of Danube at Vienna (1968) and Ulm (1980 – 1993) and of Inn (1975) at Scharding

Ganga

River Ganga originates at higher altitudes in the Himalayas and enters the plains at Hardwar. An investigation was carried out [5] to separate out the run-off component and the groundwater contribution in the 220 km sector from Hardwar and Narora (Fig 3.2).

The Himalayan precipitation run-off has a mean $\delta^{18}O$ value of about -10‰ at Hardwar. The month-wise variation of the river water $\delta^{18}O$ values at Hardwar and other stations is shown in Fig.3.3. At each station, the value reflects the net effect of the influent and effluent conditions existing between Hardwar and that station.

Rainfall in the plains has a mean $\delta^{18}O$ value of about –5.56‰, which is assumed to be the value for the local groundwater in the sector studied.

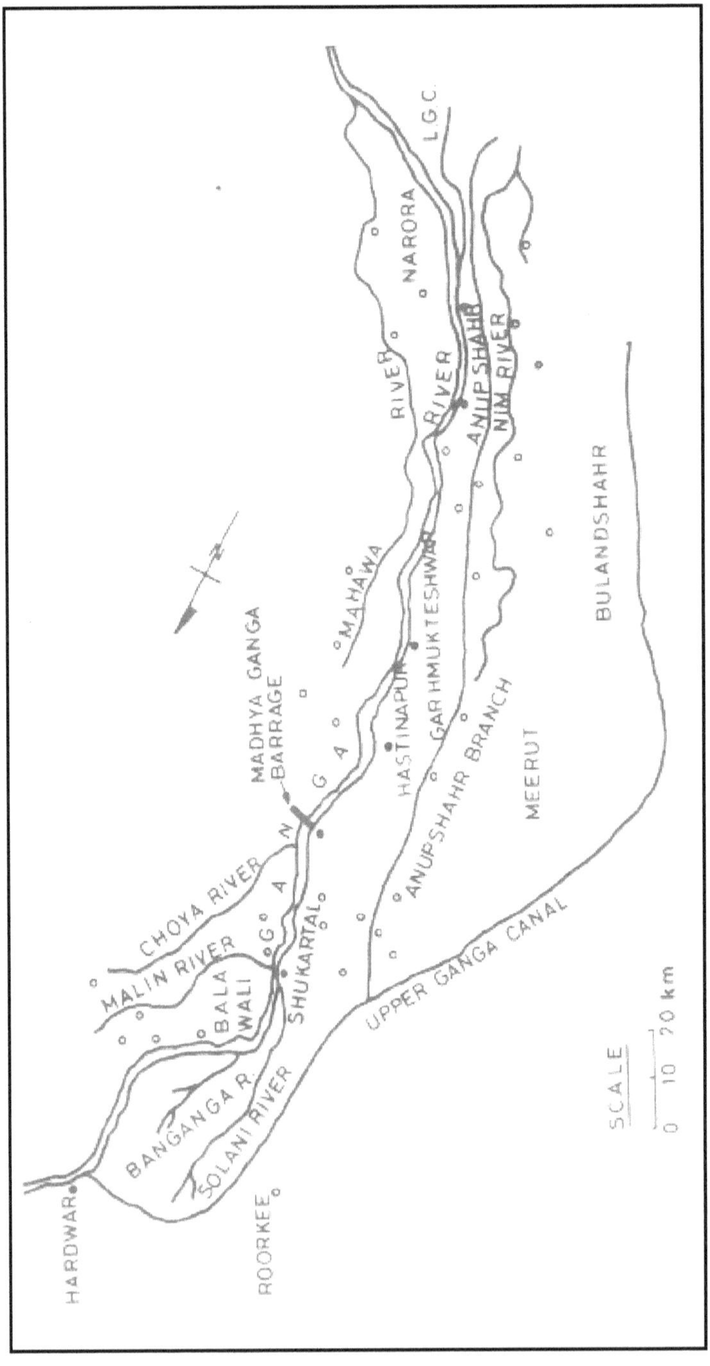

Figure 3.2 Hardwar – Narora Sector of Ganga

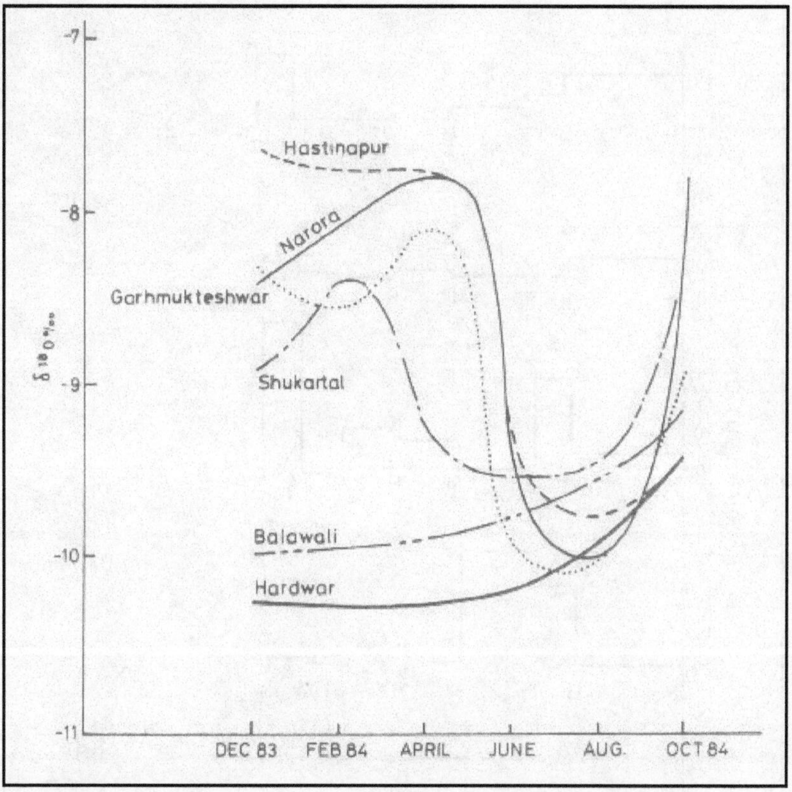

Figure 3.3 Monthly $\delta^{18}O$ variation along the Hardwar – Narora sector

Using Eq. 3.3 the groundwater contribution at each station is calculated and presented in Fig. 3.4.

It is seen that the river has hardly any groundwater contribution in August when the monsoon is at its peak. In the other months the groundwater contribution is a function of the river-groundwater level difference and the bank storage from previous effluent conditions. Groundwater contribution rises to 30 to 40% at several places during October to April.

Such a component-wise hydrographic separation of stream flow would not be easy without isotope data.

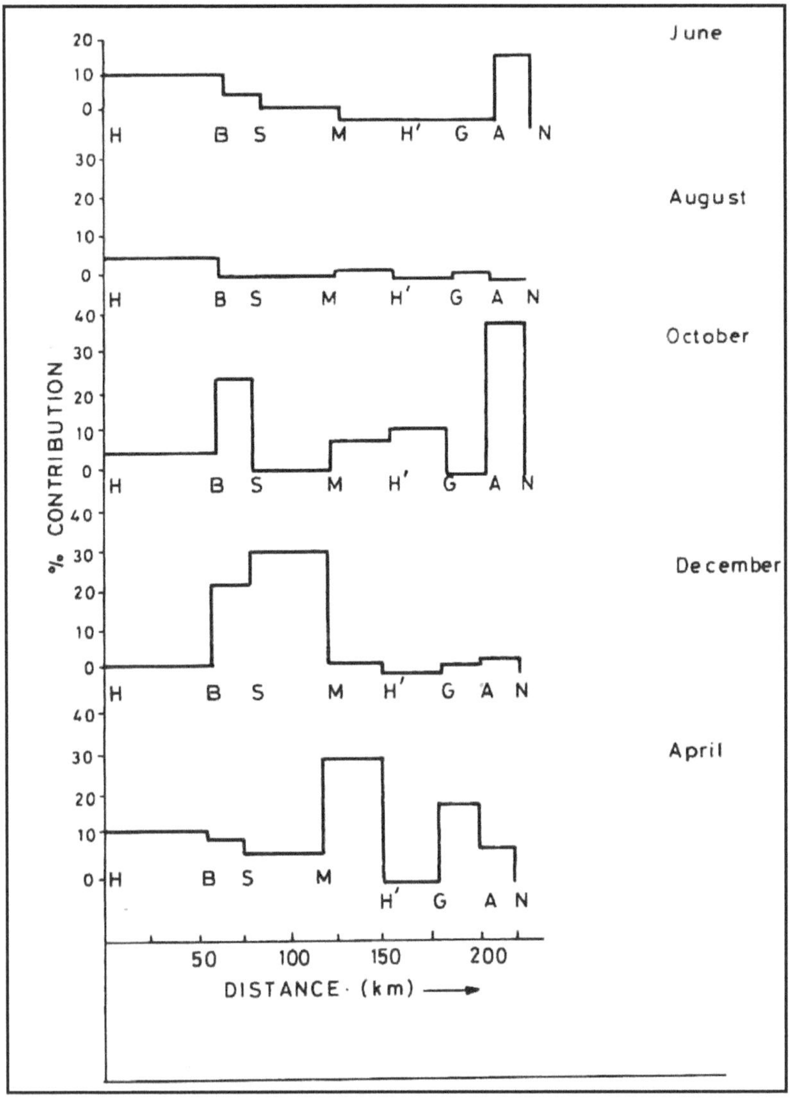

Figure 3.4 Groundwater contribution to River Ganga (H – Hardwar, B – Balawali, S – Shukartal, M – Madhya Ganga Barrage, H' – Hastinapur, G – Garhmukeshwar, A – Anupshahr, N – Narora)

3.3 Lake Dynamics

Both man-made and natural lakes are highly dependable water resources for most part of the year. Like any other water resource, optimum utilization of any lake depends on our understanding of its dynamics. The various processes of interest involved in lake dynamics are mixing, water balance and pollutant transport

3.3.1 Mixing

Mixing is a process to reduce either temperature gradient or solute concentration gradient with depth in a lake. The latter may include electrical conductivity and dissolved oxygen. To these, we can add isotope composition (δ^2H or $\delta^{18}O$). Mixing may be caused by molecular diffusion at the micro level or by turbulence at the macro level. Topography of the lake and surroundings may also contribute to mixing. When a lake is not vertically well mixed, the upper less dense and well-mixed *epilimnion* and the lower, denser *hypolimnion* are separated by an intermediate region called *metalimnion* of relatively rapid change in parameters (density, temperature). *Thermocline,* the plane of maximum temperature gradient, forms during summer and in most lakes acts as a barrier for vertical mixing.

3.3.2 Water Balance: Groundwater Inflow and Outflow Estimates

Water balance of a lake involves surface and sub-surface inflows and outflows as well as precipitation and evaporation. Usually all these have short-tem fluctuations caused by climate. In a specified time interval, the water budget of a lake has the form

$$\mathbf{(dV_L/dt) = I_S + I_G + P - O_S - O_G - E} \tag{3.4}$$

V_L is the lake volume, I_S, I_G, O_S, O_G are the surface and groundwater inflow and outflow fluxes respectively, P stands for precipitation over the lake surface and E is the evaporation flux from the lake.

The isotope tracer approach to water balance depends on the fact that the law of conservation of mass depicted in Eq. 3.4 applies to the constituent isotopes (^2H, ^{18}O) as well. We can then write the isotope balance equation as

$$\mathbf{[d(\delta_L V_L)/dt] \text{ or}}$$

$$\mathbf{\delta_L (dV_L/dt) + V_L (d\delta_L/dt) = \delta_{IS} I_S + \delta_{IG} I_G + \delta_P P - \delta_{OS} O_S - \delta_{OG} O_G - \delta_E E} \tag{3.5}$$

δ with the respective subscript represents the isotope composition of each component of the lake system.

If the variability of the components is reasonably low in the long term and the lake system is 'old' enough to be in a steady state, then

$$(dV_L/dt) = 0 \text{ and } (d\delta_L/dt) = 0 \tag{3.6}$$

We can further assume, in the case of a well-mixed lake, that the isotopic compositions of the surface and groundwater outflows are identical to the lake water i.e. $\delta_{OS} = \delta_{OG} = \delta_L$. We can then solve Eqs. 3.4 and 3.5 for the required pair of variables I_G and O_G, provided the long term averages of the other parameters are known.

$$I_G = I_S \frac{(\delta_{IS} - \delta_L)}{(\delta_L - \delta_{IG})} + P \frac{(\delta_P - \delta_L)}{(\delta_L - \delta_{IG})} + E \frac{(\delta_L - \delta_E)}{(\delta_L - \delta_{IG})} \tag{3.7}$$

$$O_G = I_S + I_G + P - E - O_S \tag{3.8}$$

The above can be further simplified if there is no surface inflow and outflow.

The most difficult parameter to estimate is the isotopic composition δ_E of the evaporation flux. It is given by the following equation derived based on Craig – Gordon model [6] of isotope fractionation during evaporation.

$$\delta_E = \frac{\alpha_{V/L}\,\delta_L - h_N\,\delta_A + \varepsilon_{V/L} + \varepsilon_{diff}}{1 - h_N - \varepsilon_{diff}} \tag{3.9}$$

where

δ_L is the isotopic composition of lake water

h_N is relative humidity of air over the lake (normalized to lake surface temperature)

δ_A is the isotopic composition of water vapour in the open atmosphere above the lake

$\alpha_{V/L}$ is the equilibrium isotope fractionation factor between vapour and liquid water at lake surface temperature

$$\varepsilon_{V/L} = \alpha_{V/L} - 1 \ (<0)$$

ε_{diff} is the diffusion fractionation = ~ - 12.5 $(1 - h_N)$‰ for ^2H
& ~ - 14.2 $(1 - h_N)$‰ for ^{18}O

Values of $\alpha_{V/L}$ for temperatures of interest can be calculated from well-established relationships [7] given below.

For ^{18}O

$$\ln {}^{18}\alpha_{V/L} = -\ln {}^{18}\alpha_{L/V} = 2.0667.10^{-3}+(0.4156/T) - (1.137.10^{3}/T^{2})$$

For ^2H

$$\ln {}^{18}\alpha_{V/L} = -\ln {}^{18}\alpha_{L/V} = -52.612.10^{-3}+(76.248/T) - (24.844.10^{3}/T^{2})$$

For a detailed discussion on the above, the reader is referred to Bibliography 2 given at the end of this chapter.

3.3.3 *Water Balance: Water Retention Time and Age of Inflowing Groundwater*

Environmental tritium offers an elegant approach [8] to estimate Water Retention Time (WRT) in the lake and also to determine the age of groundwater contributing to the lake. Rewriting the water balance equation we have

$$(dV_L/dt) = I_S+I_G+P-O_S-O_G-E \tag{3.10}$$

We can write a tritium balance equation for a lake in hydrological steady state (i.e. the inflows and outflows are balanced) as

$$V_{L.}\, dT_L/dt = T_S.I_S+T_G.I_G+T_P.P-T_L.O_S-T_L.O_G-T_E.E-T_L.\lambda.V_L \tag{3.11}$$

where T_i represents the respective tritium content of the water balance component. T_L is the tritium content of the lake and ë is the decay constant of tritium (0.05625 a^{-1}). T_E can be defined for a well-mixed lake as [9]

$$T_E = \frac{\alpha\, T_L - h\, T_A}{\alpha_K (1-h)} \tag{3.12}$$

where h is relative humidity, T_A is the tritium content of the water vapour over the lake, α is the equilibrium fractionation factor for tritium between liquid and vapour phases (~ 0.9) and α_K is the kinetic fractionation factor (~1.1). It is further assumed that the tritium content in the water vapour over the lake is in equilibrium with the local meteoric water i.e. $T_A = \alpha\, T_P$. Dividing Eq. 3.11 by V_L, we have,

$$dT_L/dt = T_S.K_{IS} + T_G.K_{IG} + T_P.K_P - T_L.K_{OS} - T_L.K_{OG} - T_E.K_E - T_L.\lambda. \quad (3.13)$$

where $K_{IS,}$ K_{IG} and K_P are the fractional contribution of each inflow component. So also are $K_{OS,}$ K_{OG} and K_E for the outflow components. For a steady state lake system, the water retention time (WRT) Γ is defined as the inverse sum of $K_{IS,}$ K_{IG} and K_P or of $K_{OS,}$ K_{OG} and K_E and has units of time.

$$\Gamma = 1/\,(K_{IS} + K_{IG} + K_P)$$

$$\text{or } 1/\,(K_{OS} + K_{OG} + K_E) \quad (3.14)$$

Assuming piston flow of groundwater contributing to the lake, we can write

$$T_{G,t} = T_{P,t-\tau}\ (e^{-\lambda\tau}) \quad (3.15)$$

where $T_{G,t}$ is the tritium content of groundwater at time t & $T_{P,t-\tau}$ is the tritium content of precipitation at time t–τ (where τ is the residence time or age of groundwater). Discretizing Eq. 3.14 in a time step of one year,

$$T_{L,t} - T_{L,t-1} = T_{P,t}K_P + T_{S,t}K_{IS} + T_{P,t-\tau}\,e^{-\lambda\tau}.K_{IG} - \frac{\alpha\, T_{L,t} - h\,\alpha T_{P,t}}{\alpha_K\ (1-h)}.\,K_E$$

$$-T_{L,t}\,K_{OG} - T_{L,t}\,K_{OS} - T_{L,t}\,\lambda \quad (3.16)$$

Assuming $T_{S,t} = T_{P,t}$ in the case of small drainage basins (i.e. the tritium level in the stream waters entering the lake retain the precipitation values) and since the lake volume remains constant, the groundwater components K_{OG} and K_{IG} can be expressed in terms of WRT (Γ) [i.e. $K_{OG} = (1/\Gamma) - K_{OS} - K_E$ and similarly $K_{IG} = (1/\Gamma) - K_{IS} - K_P$]. Hence Eq. 3.16 becomes:

$$T_{L,t}\{1 - K_E[1 - \frac{\alpha}{\alpha_K\,(1-h)}] + (1/\Gamma) + \lambda\} = T_{L,t-1} + T_{P,t}[K_P + K_{IS} +$$

$$\frac{h\alpha K_E}{\alpha_K\,(1-h)}] + T_{p,\,t-\tau-te}^{\ -\lambda\tau}[(1/\Gamma) - K_{IS} - K_P) \quad (3.17)$$

This equation 3.17 provides the required model. Tritium input function T is available from the IAEA-WMO network for most parts of the globe [10]. Using this function as well as the known values K_P, K_{IS}, K_E, h, α and α_K, the tritium values obtained for a given year for different values of WRT (Γ) and groundwater age τ are plotted. The values matching with the measured value of tritium for that year are read from the plot. The model has been applied to several lakes [8] and the model-predicted WRT and τ values are compared with the values reported in literature. Example of application of the model to the Blue Lake of Mount Gambier set in a karstic region of southeastern Australia is presented in some detail.

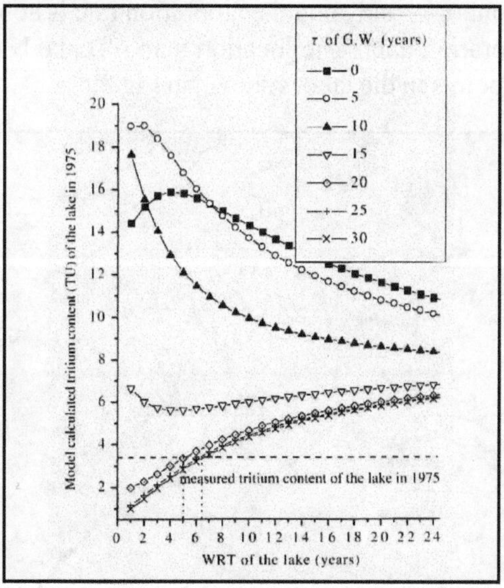

Figure 3.5 Predicted 1975 Tritium content of Blue Lake

The Blue Lake has a volume of 36 million m^3 and a surface area of about 0.6 million m^2. The recharge to the lake is entirely subterranean. Based on the reported [11] natural tritium level of 0.75 TU, a constant value of 0.75 TU was used for $T_{G,t}$ and based on the quarterly measurement of lake tritium levels, T_L for 1975 and 1976 is taken as 3.5 TU. Fig.3.5 shows the model-predicted lake tritium value for 1975 for several values of WRT and τ. It can be seen that the model gives a value of 5 – 6.5 years for WRT and a groundwater residence time of >20 years.

3.3.4 Case Study: Lake Naini [8]

Hydrodynamics

Lake Naini is situated in Lower Himalayas at an altitude of 1940 m. It has an average depth of 18.5 m (Max: 27.3 m), a surface area of 465000 m^2 and a volume of about 8.5 Mm^3. The lake is divided into two basins, Mallital (north) and Tallital (south), by a submerged transverse (east-west) ridge about 8 m below the lake surface. A catchment area of about 4.7 km^2 is situated between the altitude 1940 and 2600 m. The annual rainfall in the catchment is between 2000 and 2500 mm. Surface run-off is estimated to contribute about 45% to the lake volume. Groundwater contribution is substantial. Average relative humidity over the lake is about 64% and annual evaporation rate is about 2.6 $m.a^{-1}$ as per pan evaporation data. The location map of Lake Naini indicating the sampling points in the lake is shown in Fig.3.6.

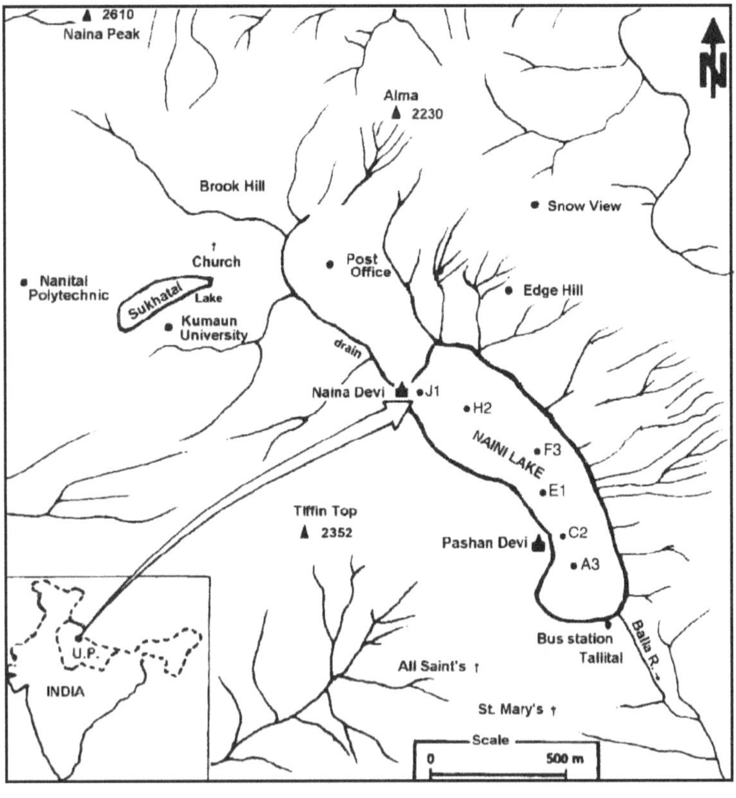

Figure 3.6 Location and drainage map of Lake Naini with sampling points

Vertical distribution of environmental isotopes [δD (or ^{2}H) and $\delta^{18}O$] and of physical parameters like temperature, electrical conductivity (EC) and dissolved oxygen, given in Figs 3.7 & 3.8 show correlative trends. The lake is stratified during March/April to October/November and well mixed thereafter. The winter overturn is in November. During the early part of the stratification, the metalimnion is at 3 m depth and at 6 m in the later part. Since the electrical conductivity is nearly constant during stratification, it is clear the stratification is temperature induced and not solute concentration induced.

Vertical distribution of all parameters shows disturbances at deeper points, probably due to groundwater inflows at those points. Most of such inferred inflows appear to be near the submerged ridge in the centre. In addition to poor vertical mixing, poor horizontal mixing is seen from the distribution of various parameters. The presence of the ridge between Mallital and Tallital further inhibits horizontal mixing and this is reflected in the profiles at deeper levels.

The isotope composition of the hypolimnion does not remain constant during stratification and the δ values of hypolimnion and epilimnion are not well correlated. These show that the retention period is short and the two regions have separate inflow systems. Variation of EC levels confirms the short retention time.

The δD - $\delta^{18}O$ plot (Fig. 3.9) of epilimnion water during stratification and of lake waters during well-mixed period show that the epilimnion waters during stratification show evaporative trend and fall on evaporation line $\delta D = 4.3\ \delta^{18}O + 18.4$. The other samples fall on the local meteoric water line (LWML) $\delta D = 7.4\delta^{18}O + 5.4$. The samples from locations C2 and E1 fall on LWML even during stratification as they are probably influenced by the groundwater inflow at those locations.

Water retention Time & groundwater age

The model described by Equation 3.17 was applied to lake Naini using the known values of K_P, K_{IS}, K_E, h, α and α_K as well of T_P. The T_P values for the period 1953 – 94 were reconstructed from IAEA-WMO data for Kabul and New Delhi and for the remaining period from tritium analyses of precipitation near the lake. The tritium levels of lake water in 1994 and 1999 (measured and averaged over stratified and well-mixed periods) were 11.2 and 11.8 TU respectively.

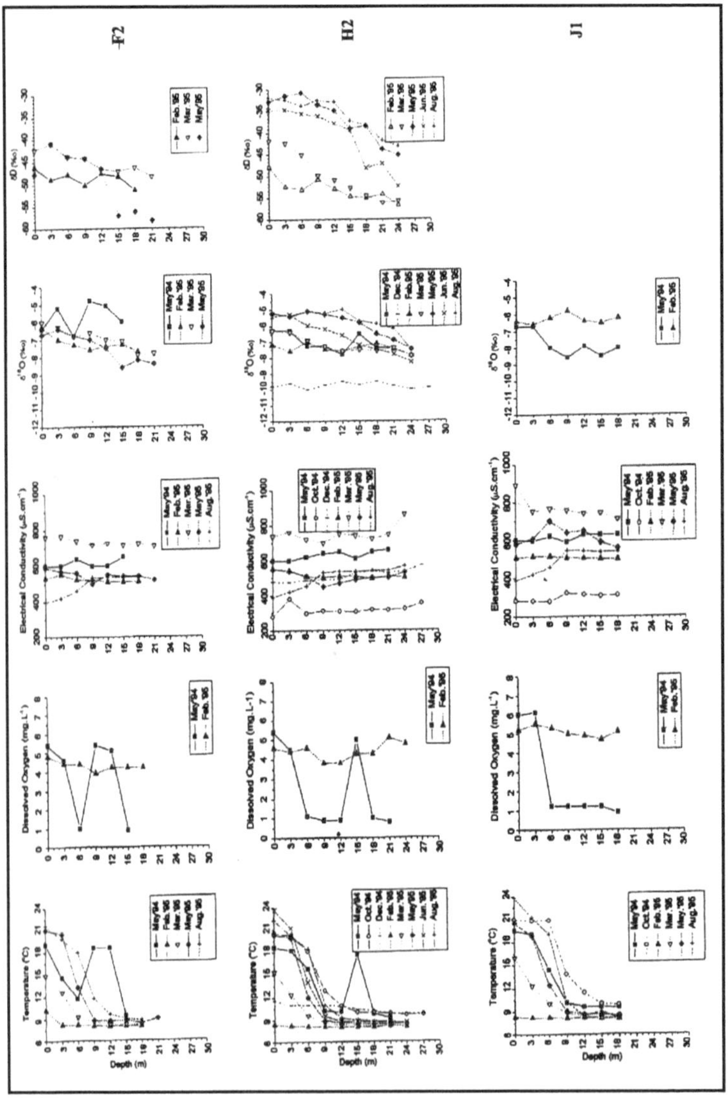

Figure 3.7 Month-wise vertical distribution of D and ^{18}O and other parameters in the northern part (sampling locations F3, H2 and J1as in Fig. III.6)

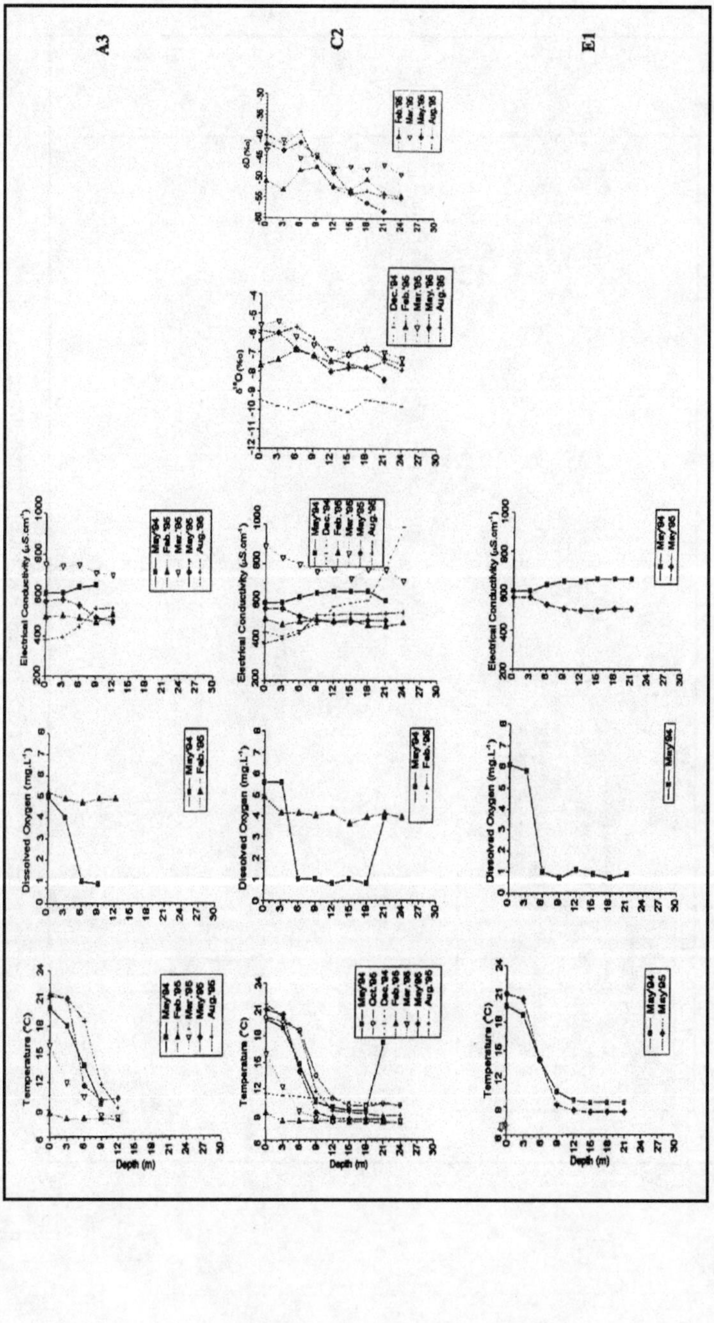

Figure 3.8 Month-wise vertical distribution of D and ^{18}O and other parameters in the southern part (sampling locations A3, C2 and E1 as in Fig. III.6)

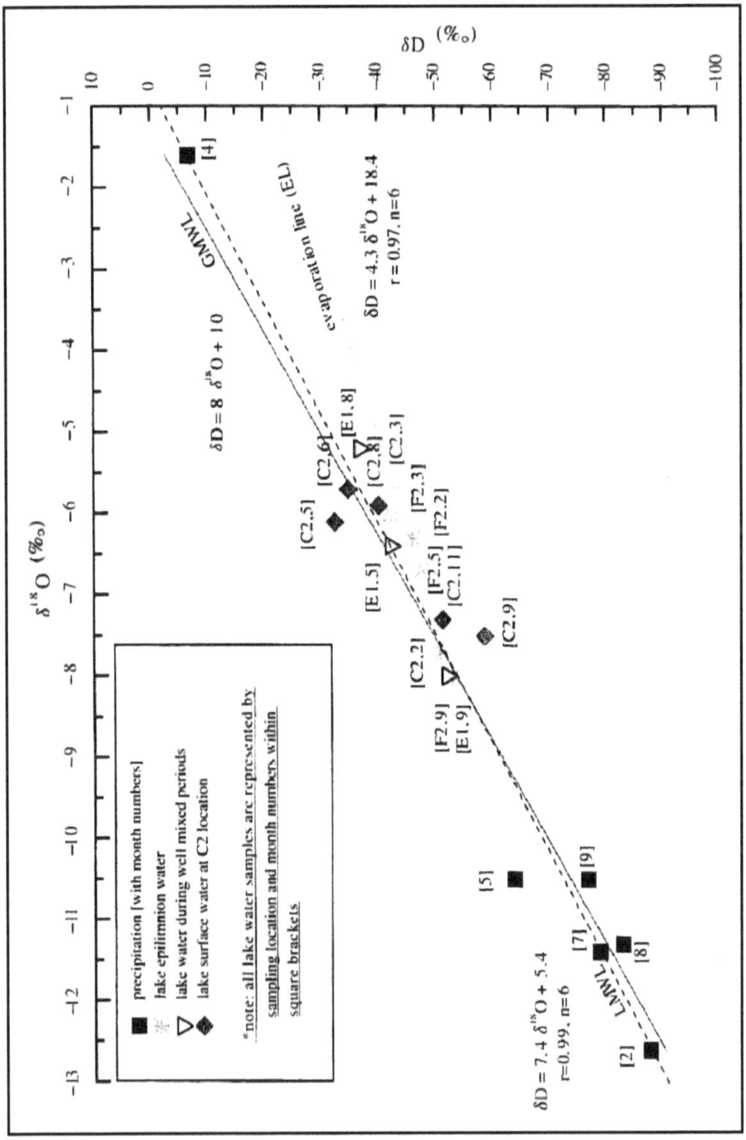

Figure 3.9. δD – $\delta^{18}O$ plot of Lake Naini

The model-calculated lake tritium values for 1999 for different values of WRT (< terminal value) and groundwater age τ are plotted in Fig. 3.10.

It is seen that the measured tritium value of 11.8 TU for 1999 intersects all curves (τ > 0) at WRT ~ 2 years and also the curve τ = 15 years at MRT ~ 6 years. The calculated lake tritium levels for several years were recalculated for various combinations of WRT (2 and 6 years) and τ > 0 and plotted in Fig.3.11.

It is seen that the measured lake tritium levels for 1994 and 1999 are best predicted using WRT of 2 years and groundwater age of 14 years.

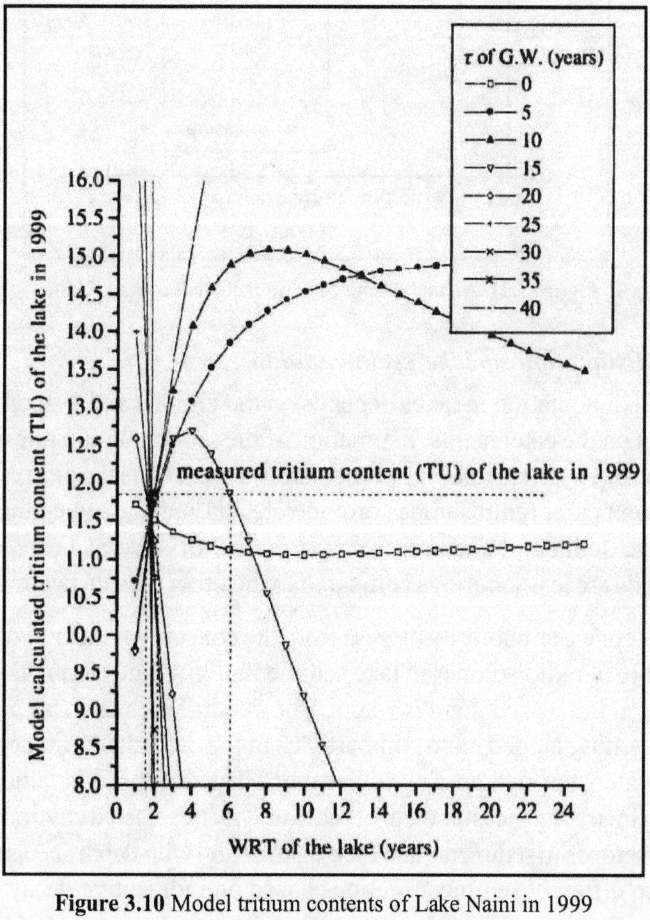

Figure 3.10 Model tritium contents of Lake Naini in 1999

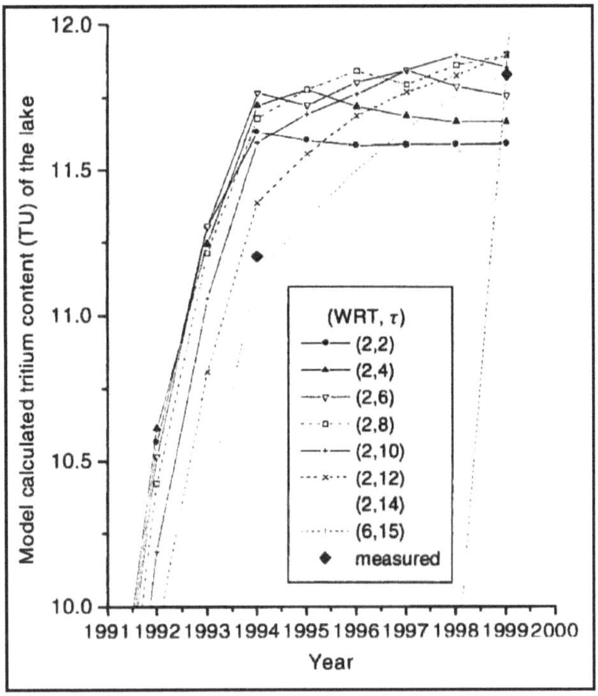

Figure 3.11 Annual Predicted tritium content of Lake Naini

3.3.5 Estimation of lake sedimentation rates

Sedimentation in lakes depends on the climate and also on man's impact on the catchments. Estimation of the sedimentation rates, in the long term, is essential for the proper management of a lake or a reservoir. Environmental radioisotopes provide elegant methods to estimate the average sedimentation rates over a period of several years. These methods are less laborious compared to the other established methods.

Here we temporarily digress from the concept of tracer hydrology. The role of radioisotopes in lake sedimentation studies is not as tracers of water, but as sediment markers. They enter lake waters as aerosols or by getting tagged on to soil particles in the catchments and entering lake waters through precipitation runoff. The technique is radiometric dating involving measurement of radioactivity of selected environmental radioisotopes in sediments and their distribution with depth and assigning ages to different sediment sections based on radioactive decay.

Two radioisotopes have been found to be very useful for the estimation of lake sedimentation rates. They are ^{210}Pb (lead-210) and ^{137}Cs.

^{210}Pb method

^{210}Pb is a member of the natural uranium family (^{238}U decay series, See Fig. 1.2). Let's rewrite the series, in a simplified way, starting from Radium-226 (^{226}Ra).

$$\mathbf{^{226}Ra \rightarrow \alpha \rightarrow {}^{222}Rn \rightarrow \alpha \rightarrow {}^{218}Po \rightarrow \alpha \rightarrow {}^{214}Pb \rightarrow \beta \rightarrow {}^{214}Bi \rightarrow \beta \rightarrow {}^{214}Po}$$

(3.8 d) (3 min) (27 min) (<1 ms) ↓

$$\mathbf{(Stable)\ {}^{206}Pb \leftarrow \alpha \leftarrow {}^{210}Po \leftarrow \beta \leftarrow {}^{210}Bi \leftarrow \beta \leftarrow {}^{210}Pb \leftarrow \alpha \leftarrow} \downarrow$$

(138 d) (5 d) (22.3 a)

α and β refer to the respective decay emissions and the half-lives of radioisotopes are given in parentheses.

Radon-222 (^{222}Rn) being a gas partially escapes into the atmosphere from the lithosphere where it is produced. It decays with a half-life of less than four days followed by short-lived products till ^{210}Pb with a half-life of 22.3 years. ^{210}Pb and other short-lived daughter products are removed through aerosols by dry fallout and precipitation.

The residence time of ^{210}Pb in the atmosphere is about 10 days. The ^{210}Pb flux from atmosphere to land and water surfaces is a continuous process. This fallout was generally found to be 100-150 Bq/ m^2/a near European lakes. It is removed rapidly by suspended particulate matter and accumulates in lake sediments. This ^{210}Pb is termed as 'unsupported ^{210}Pb ' or '$^{210}Pb_{excess}$'. It is because this ^{210}Pb is a consequence of radon, which escaped from its parent ^{226}Ra. But, the uranium family naturally present in lake sediments continues to produce what is called 'supported ^{210}Pb'. 'Supported', because it is in 'secular equilibrium' with ^{226}Ra in the sediment. Measurement of ^{226}Ra will help to determine the 'fallout ^{210}Pb' or '$^{210}Pb_{excess}$' in the following way.

In any given sediment section

$^{210}Pb_{excess}$ = Total ^{210}Pb - ^{226}Ra activity

How does this '$^{210}Pb_{excess}$' relate to sedimentation rate?

A typical semi-log plot (Fig 3.12) of both total ^{210}Pb and $^{210}Pb_{excess}$ vs. depth illustrates [12] how sedimentation rates are calculated and ages assigned to a sediment profile. The low and uniform level of radioactivity below 23 cm represents supported ^{210}Pb where $^{210}Pb_{excess}$ decayed to an insignificant level. This supported ^{210}Pb level (0.6 dpm/g) is subtracted from the total activity at each sampling depth and the resulting $^{210}Pb_{excess}$ level is plotted (solid line). This is uniformly high over the top 9 cm and decreases with constant slope between 10 and 23 cm. The uniform level in the surface layer is attributed to mixing of the sediment by organisms (bioturbation) and physical processes. This situation may change from one lake to another.

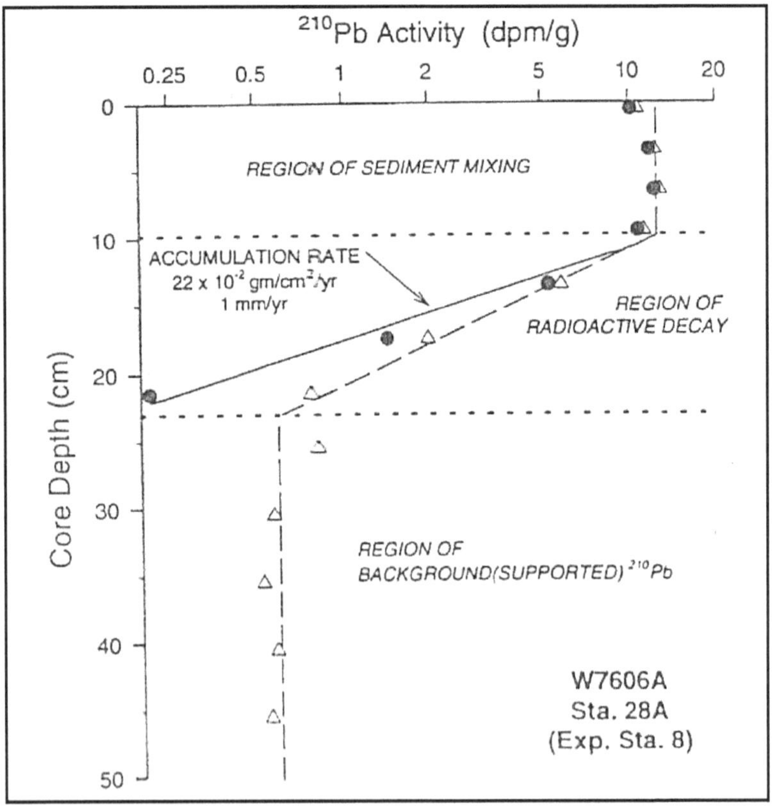

Figure 3.12 A typical profile of ^{210}Pb activity in a sediment core (Dashed line represents total ^{210}Pb and solid line is least square regression through $^{210}Pb_{excess}$ shown as solid circles)

The sedimentation can be calculated from the slope of decreasing concentration from

$$A(z) = A(0)\, e^{-bz}$$

Where A (z) is $^{210}Pb_{excess}$ activity at depth z, A (0) is the activity at the surface (or below the uniformly mixed layer), b is the slope defined by regression through the data. Changing the units of slope from depth to time, the rate of sedimentation S is given by

$$S = (\lambda/b)\ \text{cm/a}$$

where λ is the decay constant of ^{210}Pb (0.31/a)

The sedimentation rate can be converted to units of g/cm²/a when the grain density and water content are known.

^{137}Cs method

Main contributor of ^{137}Cs (Cesium-137) to the atmosphere was atmospheric testing of thermonuclear weapons in late fifties and early sixties with maximum input to the stratosphere in 1962/63. It had been coming down to the geosphere by precipitation and getting rapidly and strongly adsorbed on to fine soil particles. The bomb fallout ^{137}Cs levels were estimated to be around 1-3 Bq/m²/a. There was higher fallout due to the Chernobyl nuclear reactor accident in 1986, but the deposition was constrained to Europe. ^{137}C is a gamma emitter with energy of 0.66 MeV and a half-life of 30 years. Because of the long half-life the radioisotope is still detectable in lake sediments and hence useful for estimation of sedimentation rates in the last forty years or so.

The soil particles strongly labeled with ^{137}Cs in lake catchments find their way to the lake through run off and get accumulated in sediments. The procedure to apply ^{137}Cs for estimation of sedimentation rates is to identify the position of the 1963 fallout peak in a sediment core. The amount of sediment accumulated above the peak divided by the number of years since 1963 will yield sedimentation rate.

Radiometric techniques for ^{210}Pb & ^{137}Cs

The radiometric technique starts with collection of undisturbed cores from the lakebed using a suitable corer. The core is sliced into suitable sections (1-2 cm) for radiometric analysis.

Total ^{210}Pb is assayed by measuring the 5.3 MeV alphas emitted by its granddaughter ^{210}Po, which is assumed to be in secular equilibrium with ^{210}Pb. Special radiochemical procedures [13] are adopted to extract polonium from sediment and deposit polonium nuclides on silver planchettes for α counting with a solid state nuclear detector. The level of supported ^{210}Pb is measured by direct gamma counting of Radium-226 with which the former is expected to be in secular equilibrium. ^{137}Cs, being a hard gamma emitter can be directly measured in a well-calibrated low background NaI scintillation counter or in a hyper pure Germanium solid-state detector.

3.3.5.1 Case Study: Lake Naini [13]

The methods described above have been successfully applied to several lakes around the world including Lake Naini in Uttaranchal, India.

A general description of the lake is given earlier in 3.3.4. Fig. 3.13 gives the locations (Q, V & S) from where sample cores were drawn in Lake Naini. A gravity corer with inner and outer diameter of 5.2 & 6.0 cm respectively was used. The methods explained above were used for ^{210}Pb and ^{137}Cs analyses.

The depth profiles of total ^{210}Pb and ^{137}Cs are given in Fig. 3.14.

The following table gives the estimated sedimentation rates obtained with ^{210}Pb and ^{137}Cs methods.

Sampling station (water depth in m)	Average sedimentation rate (with ^{210}Pb method)		Average sedimentation rate (with ^{137}Cs method)	
	Linear (cm/a)	Mass (g/cm^2/a)	Linear (cm/a)	Mass (g/cm^2/a)
V (24)	0.5	0.11	0.6	0.14
Q (20)	0.6	0.15	0.7	0.17
S (20)	1.2	0.29	1.4	0.32

In Fig. 3.14 it can be seen that at location V the ^{210}Pb profile showed a near exponential decrease of concentration with depth reaching a constant value maintained by the decay of ^{226}Ra. But in other locations the profiles did not approximate to an exponential one. The corresponding measurements on ^{137}Cs did not indicate any loss of top portions during coring. After a detailed analysis of all possible causes such as slumping, sediment focusing, bioturbation etc. it was concluded [13] that the cause for the non-monotonic profiles at Q and S could be the actual change in sedimentation rate with varying sediment composition and dissimilar histories of deposition.

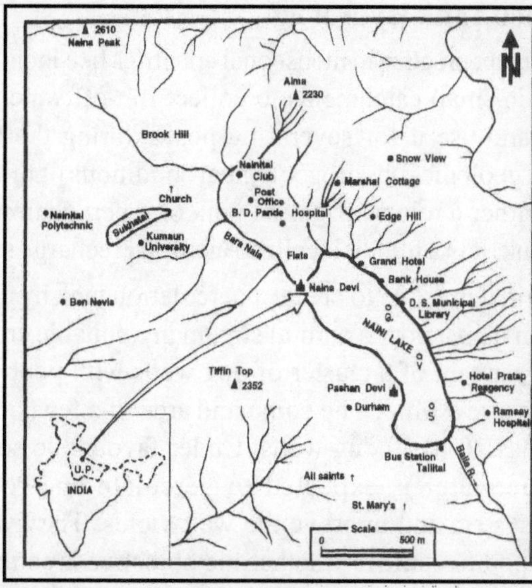

Figure 3.13 Location map of Lake Naini (Open circles indicate core-sampling locations Q, V & S)

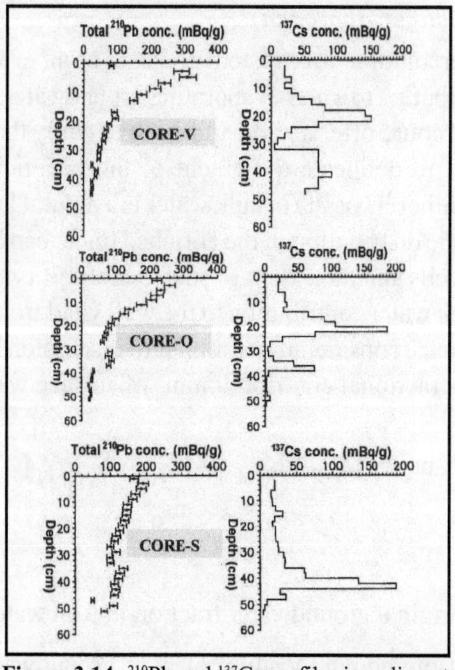

Figure 3.14 ^{210}Pb and ^{137}Cs profiles in sediment cores from 3 locations in Lake Naini

3.4 Percolation Tank Hydrology

It is an ancient practice in monsoonal countries like India to construct storage tanks in small catchments to collect run-off water during the rainy season and use it for several purposes during the dry season. Depending on geological/hydrogeological conditions of the terrain, the tank may be either a minor irrigation tank or a percolation tank. As a percolation tank, it is a highly localized artificial recharge system.

It is a normal practice to create a percolation tank by constructing a small earthen dam across a natural stream in a suitable cross-section. It is usually upstream of a cluster of dug wells with poor yields. The tank volume is a few Mm^3. The command area of a few km^2 under the tank usually has 20 to 50 dug wells. Under favourable soil and rock conditions the water is expected to percolate and recharge the groundwater system and improve the well yields. The hydrologist is often called upon to confirm whether the objective is achieved and to quantify the benefit, wherever possible.

3.4.1 Stable Isotope Approach

Since a percolation tank system consists of an evaporating body (the tank) contributing to a non-evaporating groundwater system, a sable isotope method could offer a good approach to study the effectiveness of the tank and to delineate the extent of the beneficial effect. The enrichment of either ^2H or ^{18}O in tank water is a natural labeling process and a study on the distribution of the enriched tracer can help to identify the benefited wells and the extent of such benefit. It is also possible to estimate the tank water contribution to the well yield from mass balance and isotope balance considerations similar to discussion in 3.2. It can be shown that the fractional contribution m_S of surface water to the well water is

$$\begin{aligned} \mathbf{m_S} &= [\delta_{PG} - \delta_{G\,(d,\,t)}] / [\delta_{PG} - \delta_{S\,(t-t')}] \\ &= \mathbf{1 - m_G} \end{aligned} \tag{3.18}$$

where

m_G is the original groundwater fraction in well water

δ_{PG} is the isotopic composition of pure groundwater beyond the recharge zone of the tank

$\delta_{G\,(d,\,t)}$ is the isotopic composition of the admixture at a distance d from the tank and at time t

$\delta_{S\,(t-t')}$ is the isotopic composition of the tank water at time t having traveled a distance d.

t' is the average travel time of percolating water to reach distance d.

The above treatment assumes well-mixed tank and wells. This is generally true as the tanks are shallow and the wells are continuously exploited. The applicability of the method is based on the following considerations.

- Highly localized nature of the problem
- Short duration of the study ensuring no large seasonal variation in isotopic compositions of the components
- No addition to the tank during investigation
- No other known source of recharge to groundwater during the period of study

3.4.2 Case Study: Shindawane Percolation Tank

This tank situated near Pune (Poona) in India is shallow with an average depth of 3-4 m and impounds about 0.5 to 1 Mm^3 of water during the June – September monsoon season. Fig. 3.15 shows the plan of the tank and the command area under irrigation. The command area, which is about 2.5 to 3 km^2, extends up to 2 km from the dam. The dug wells (about 40) penetrate soft rock overburden up to the water table and then fissured hard rock, which is exposed at the tank bed. The percolating medium is thus fissured hard rock with jointed planes and fractures. An additional hydrological feature of the system is the unlined irrigation canal (Mutha R.B. canal) flowing at the southern edge of the command area. It has an average width of 16 m and carries a discharge of 14 m^3 /s.

In a study [14] on the efficacy of the tank, ^2H (D) analyses of tank water, canal water and the well waters were carried out twice in the dry season; once in December and the second time in March.

During this period, δD of tank water was enriched from -41‰ to -19‰, whereas the canal water originating from a large reservoir (Mutha) got enriched from -19‰ to -9‰. In both cases the enrichment percentage (ΔδD%) was similar, 53 to 54%. The well waters, which had δD varying from-25‰ to -57‰ (based on their past history in a heterogeneous hard rock system) all got enriched to varying percentage levels (ΔδD% = 17 to 80%). The considerable difference in D-labeling of the two surface bodies enabled identification of wells recharged mainly by the tank, by the canal or partially by both. Fig.3.16 shows the variation of ΔδD% with distance from the tank along a selected traverse.

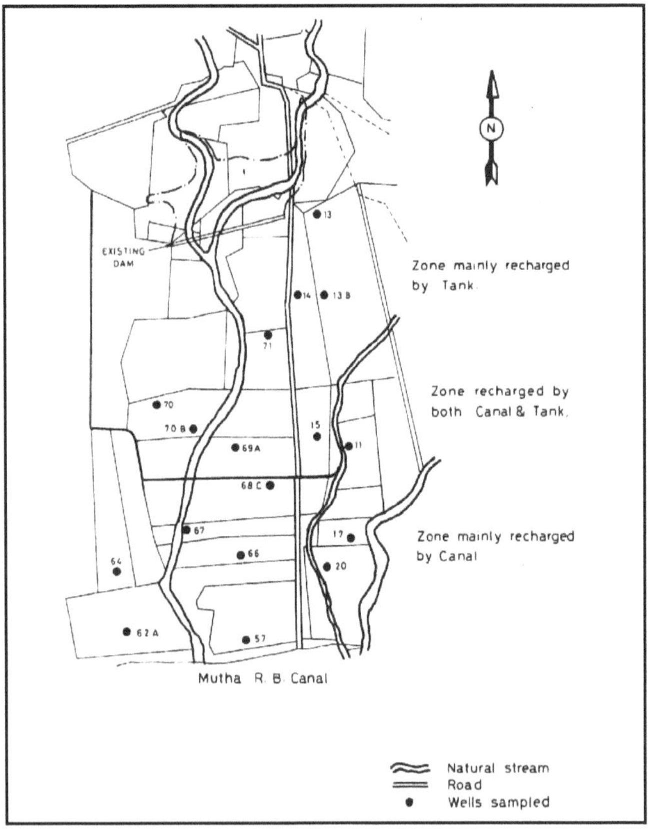

Figure 3.15 Shindawane percolation tank - Layout

Equation (3.18) cannot be easily applied in the present case due to the presence of the unlined canal in the system.

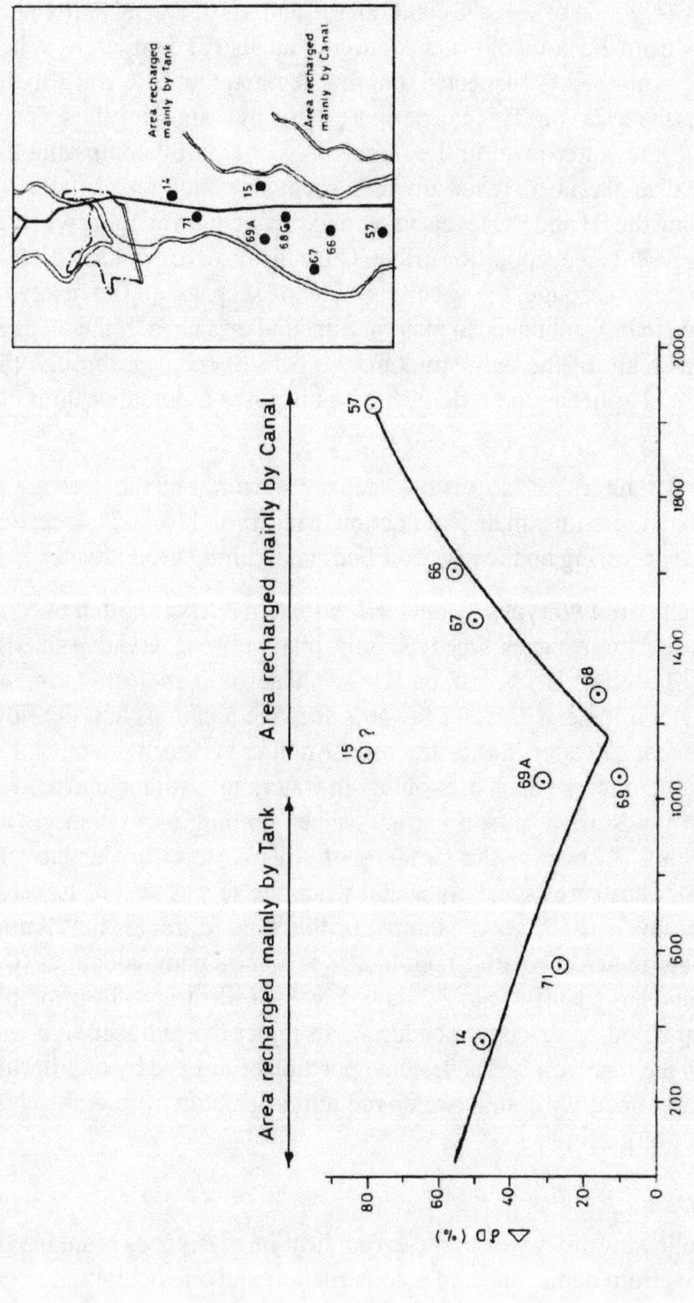

Figure 3.16 ΔδD% vs. distance from tank

3.5 Seepages from Dams and Reservoirs

Investigation of seepage conditions in dams is of crucial importance not only from the point of water economy, but also of dam safety. When seepage is noticed or suspected, the first requirement is to confirm the source is the reservoir. This can be done; firstly by comparing the seepage rate with the water level in the reservoir, secondly by comparing the chemical analysis of reservoir and seepage water and finally by measuring the ^{2}H and ^{18}O levels in seepage water and verifying whether they represent an evaporation effect. Once the reservoir is identified as the source of seepage, the specific points of seepage on the reservoir side need to be established to plan for remedial measures. If the seepage is downstream of the dam structure the path of seepage through the structure also needs to be delineated as it poses a potential threat to dam safety.

In certain cases, no visible seepage occurs, but the seepage is suspected due to unexplained reduction in reservoir levels. The seepage could be occurring on the reservoir beds and joining groundwater.

There are two types of injected radioactive tracers often used in studies on dam seepages. One type is the relatively conservative anionic tracer like $NH_4{}^{82}Br$, $K^{82}Br$ or $K^{131}I$. These help in following the movement of reservoir water towards seepage locations and the flow of the seepage water in the structure. Another type of tracers, called adsorbable tracers which are soluble in water, but cationic and tend to get easily adsorbed on soil surfaces when coming into contact with them. This is because the surfaces of soil particles are negatively charged. These are especially useful when the seepages are suspected on reservoir beds. A good example of this type of tracers is **^{198}Au as chloroauric acid ($HAuCl_4$)**. Gold-198 is a good gamma emitter (0.4 MeV) and has a Half-life of 2.3 days. Fig. 3.17 illustrates the principle of the method [15]. The procedure is to inject the radioactive tracer close to the reservoir bed at selected positions followed by monitoring the bed surface with underwater radiation detectors, preferably NaI scintillation detectors.

3.5.1 Dams

Following are two examples of application of isotopes in studies on seepages from dams: one under construction and one operating.

Srisailam

This major masonry-concrete dam was under construction in the late sixties in Andhra Pradesh, India. To facilitate its construction, two cofferdams were constructed and the inundated water between the two could not be emptied due to suspected seepages from the downstream cofferdam. Grouting was not effective, as the seepage entry points and flow path were not clearly known. A radiotracer investigation was carried out using ^{82}Br and ^{131}I, both as potassium salts [16]. Injecting the tracers at selected injection sites and monitoring the impounded water and the available boreholes (made for grouting) in the body of the dam, possible entry points and the flow paths through the dam could be delineated (Fig.3.18)

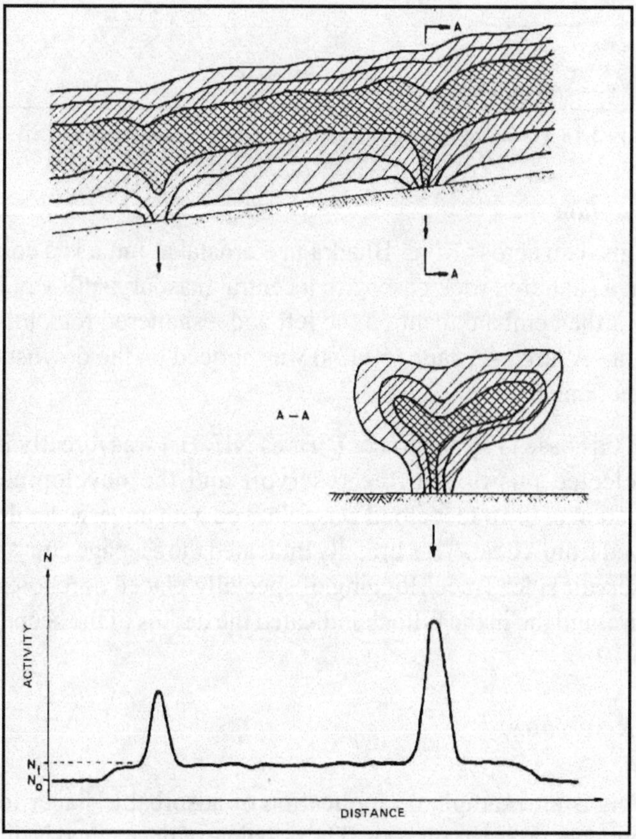

Figure 3.17 Principle of use of adsorbable radiotracer to locate seepage points on reservoir beds

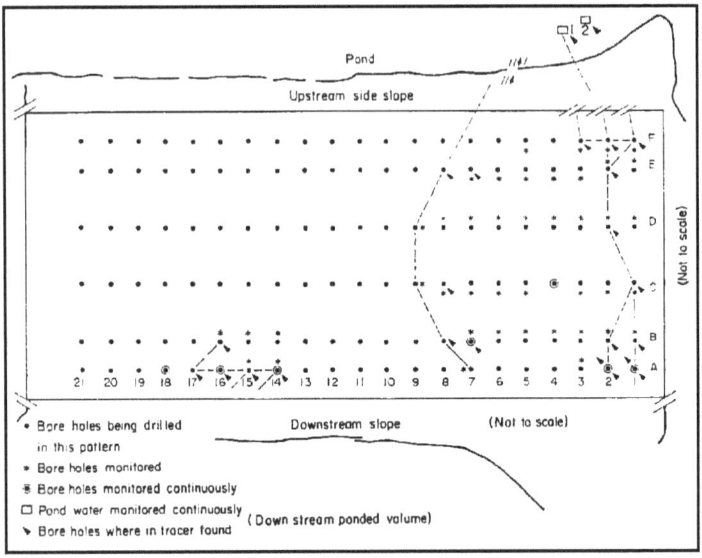

Figure 3.18 Tracer transport paths in the downstream cofferdam at Srisailam

Bhadra dam

This dam across River Bhadra in Karnataka, India is a composite dam on a shattered rock base with a central masonry spillway flanked by an earthen embankment on the left and a shattered rock hillock on the right. A large seepage (60 L/s) was noticed on the downstream of the right flank.

In this case [17], the tracer (^{82}Br as NH_4 Br) was directly injected at a selected position in the reservoir and the development and propagation of the tracer cloud was followed with underwater detectors deployed from a boat. This broadly indicated the seepage entry zone on the reservoir side. Tracer logging of specially placed boreholes on the shoreline and the in the hillock indicated the depths of the seepage path (Fig. 3.19).

3.5.2 Reservoirs

Borda

This is an example of application of adsorbable tracer to locate seepage zones on reservoir beds. The principle of the method is illustrated in Fig.3.17.

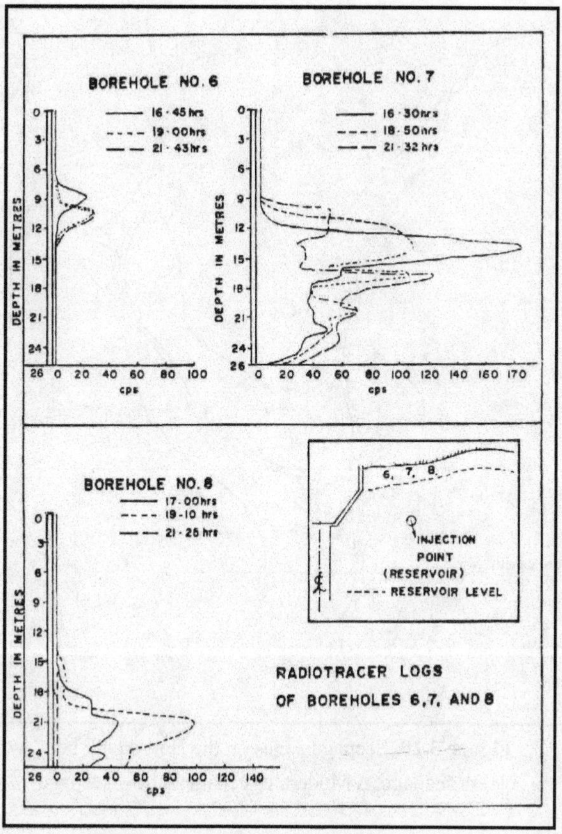

Figure 3.19 Tracer logs from boreholes

A minor irrigation tank constructed in the limestone terrain at Borda in Yeotmal district of Maharashtra, India could not retain water for long after monsoon due to heavy seepage on the reservoir bed. In a study [18] to locate the seepage zones on the reservoir bed, four aqueous tracer injections of radioactive gold-198 as $HAuCl_4$ were made into the reservoir from different locations at a predetermined flow rate. As mentioned earlier, gold-198 was chosen as the cation would move along the flow lines caused by the seepage and get adsorbed on the bed at the seepage zones. The adsorbed tracer positions were then located using underwater nuclear detectors towed from a boat. Though the results showed general seepage on limestone bed of the reservoir, particularly high seepage zones could be identified (Fig. 3.20). The high seepage zones were mainly along the two erstwhile streambeds.

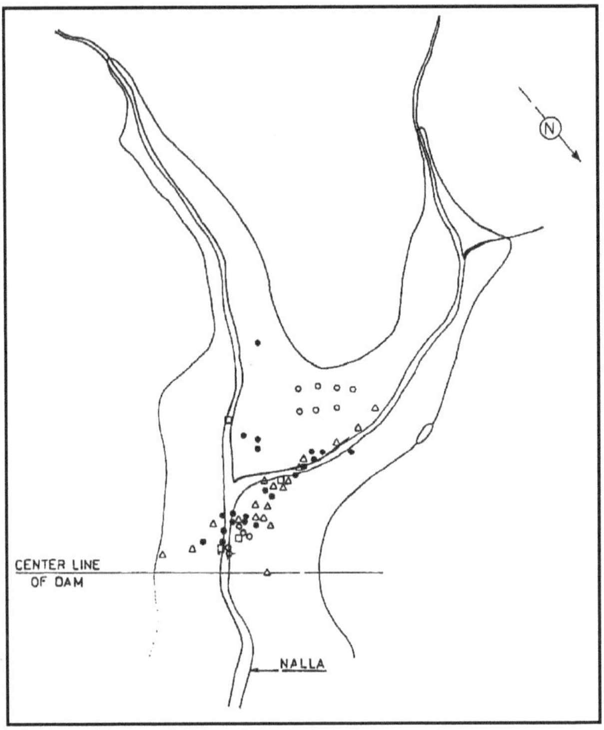

Figure 3.20 Seepage zones on the Borda tank bed
• High seepage, Δ Moderate seepage, o Low seepage

Samuel Reservoir, Brazil

Since a reservoir is an evaporating body and evaporation causes enrichment and provides a specific signature on the $\delta^2H - \delta\,^{18}O$ plot (see 2.3.1.2), it is possible to relate a suspected seepage to reservoir.

An example of a stable isotope application is given in Fig.3.21. In a study carried out to identify the origin of the seepage under a dyke of Samuel Reservoir, Porto Velho, Brazil, stable isotope compositions were used besides fluorescent dyes and artificial tritium [19]. As the reservoir volume fluctuated widely from rainy season to dry period, there was systematic variation of the isotopic composition. This variation could be traced to a number of observation wells behind the dyke, thus providing information not only on the identity of source of seepage, but also on the approximate transit times of water from the reservoir to the observation wells.

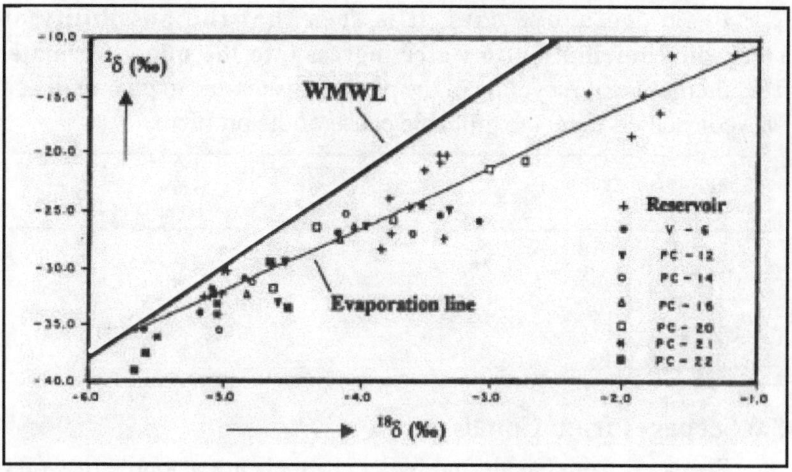

Figure 3.21 $\delta\,^{2}H - \delta^{18}O$ plot of wells behind the dyke (v and P.C's represent the wells)

Reservoir seepage into Mine Waters

The open cast Jhamrakotra rock phosphate mine near Udaipur city in Rajasthan, India is an open cast mine and spread over 20 km^2. The mine was experiencing severe water ingress in 1994 when the mining activity reached the groundwater table. The main aquifer is dolomitic limestone with solution cavities of 1 to 9 m in size. Pumping out at a rate of 1000 to 1200m^3 /h from a battery of tube wells did not significantly help. A few man-made reservoirs exist within a radius of 9 km; among them are Udaya Sagar Lake at 9 km, Jhamri Reservoir at 5 km and Jaisamand Lake. These were suspected to be possible sources as they could contribute to the mine water through cavernous limestone. An isotope investigation [20] was carried out to investigate this possibility. The investigation yielded the following data in the month of July 1995.

From the above table, it is evident the tube well water in the mine area is different from the water of any reservoir. The mine water appears to represent local precipitation whereas the reservoir waters all show evaporative enrichment in stable isotopes. Tritium values of mine water indicate some mixing of modern water with comparatively older water. This could be due to continuous pumping from different lithological units and probably recycling due to improper drainage of pumped water.

From the above data, it is seen that the possibility of reservoir contribution to water ingress into the mine is remote. The alternative of recycling of the pumped water due to poor drainage was considered to be the probable cause of the problem.

Sample Source	$\delta^{18}O$ (‰)	^3H (TU)
Tube Wells in Mine Area (Dolomite)	-4.1 to –4.7	2.8 to 5
Jhamri Reservoir	-0.2	7
Udaya Sagar Lake	+2.2	10
Jaisamand Lake	+1.2	9

3.6 Seepages from Canals

Seepages from canals and hydel channels are a source of worry to water resources managers as the unwelcome effects are many; loss of precious irrigation water, waterlogging resulting in agricultural problems and expensive repairs like lining the canals. There are two aspects to the seepage problem. One is confirmation of the existence of seepage and the second is the identification of major seepage zones through estimation of seepage losses.

3.6.1 Canal Seepage Identification, Case of Chilla Hydel Channel

When a canal system is commissioned, there is a possibility of a rise in the local groundwater table, which might over a period of time cause water logging. This rise could be either due to seepage of canal water or the result of the hydrostatic pressure exerted by the canal system on shallow groundwater table. It is possible to verify the cause if the canal water and the groundwater have different origins and consequently have different chemical and/or isotopic composition (^2H & ^{18}O).

The stable isotope method was applied [21] to Chilla Hydel Channel in Uttaranchal, India. The channel, which has a bed width of 12.5 m and a depth of over 9 m, starts from Virabhadra barrage near Rishikesh and reaches the Chilla Power House, 14 km downstream. (Fig.3.22). The channel runs along the left edge of Ganga valley and carries a discharge of 600 m^3/s. The channel is entirely lined with concrete tiles.

When the channel was commissioned in 1980, several springs appeared on the right bank at a distance of 7.5 to 9 km from the barrage.

The channel was suspected to have developed seepages in spite of the lining. To verify the seepage condition, the stable isotope method was applied.

The following table shows the isotope composition in two sets of samples collected in two different seasons and also electrical conductivity in the second set.

It can be seen that the channel water originating at a higher altitude in the Himalayas is isotopically lighter and the spring water and local groundwater are isotopically similar. The difference in the conductivity levels confirms that the springs cannot be due to seepages from the channel. The alternative explanation is the reduction of ground level

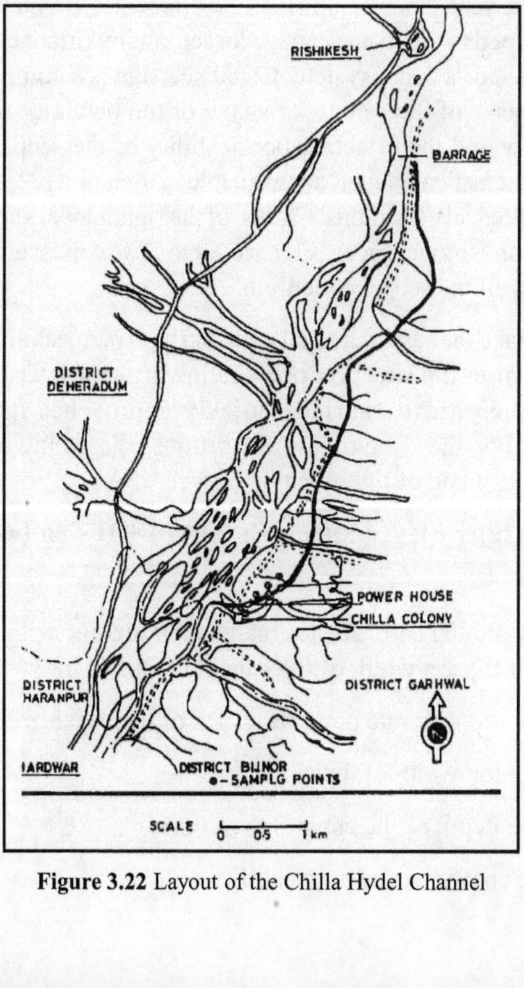

Figure 3.22 Layout of the Chilla Hydel Channel

during channel construction coupled with the hydrostatic pressure of the channel itself caused springing up of the groundwater.

S.No	Location	δ^2H (‰)	EC (μs/cm)
1	Channel	-72	-
	Springs	-55	-
	Groundwater	-54	-
2	Channel	-58	110
	Springs	-49	450

3.6.2 Canal Seepage Estimation, Case of Ganga Canal

Over the years, many empirical and direct measurement techniques were developed to estimate seepage losses, an important component of water balance of a canal system. Canal seepage is a complex function of the geometry of the canal, the shape of the phreatic surface of the seepage flow and the effective permeability of the seepage medium. Various mathematical models are available in literature [22] for a number of hydrogeological conditions. Choice of the right model appears crucial. The role of an isotope tracer is restricted to *in situ* measurement of the permeability of the seepage medium.

Let us take the case of a canal with "no high permeability underlying layer" similar to the situation of a shallow groundwater table. In this case [23] the phreatic surface quickly approaches its horizontal asymptote. For these boundary conditions, the following equation describes the shape of the phreatic surface

$$\mathbf{x = (q/2k) + [(B - q/k)/2] \cosh (\pi ky/q) + H \sinh (\pi ky/q)} \quad (3.19)$$

where

x and y are the co-ordinates on the phreatic line with the origin at the centre of the top width of the canal (x = 0.5 B for y =0)

q is the seepage rate per unit length of the canal

B is the top width of the canal

H is the depth of the canal

k is the coefficient of permeability

For the determination of q in the above equation, the only requirement is the field measurement of the coefficient of permeability. k is determined by measuring the filtration velocity v_f of the phreatic surface using the radiotracer method described in 2.2.1.3 and using the relationship

$k = (v_f / I)$ where I is the hydraulic gradient.

The above approach was applied to the unlined Ganga canal near Roorkee in Northern India [22]. The canal has trapezoidal cross-section (Fig.3.23) and is cut through gangetic alluvium. It has a top width of ~70 m, a maximum depth of over 4 m and carries a discharge of about 200 m^3 /s.

Several hand auger boreholes were made on either bank of the canal. Using ^{82}Br and ^{131}I as injected tracers in the boreholes, filtration velocities were determined. The average values were 1 cm/d on the right bank and 1.5 cm/d on the left bank leading to k values of 0.6 m/d and 0.5 m/d respectively. Using these values the phreatic surface on each bank was computer simulated (Eq. 3.19) for different values of q/2 and compared it with the observed phreatic surface. The q/2 value corresponding to the best fit was chosen for each bank. The total seepage rate (for both the banks) was 2.2 m^3/day /meter length of canal.

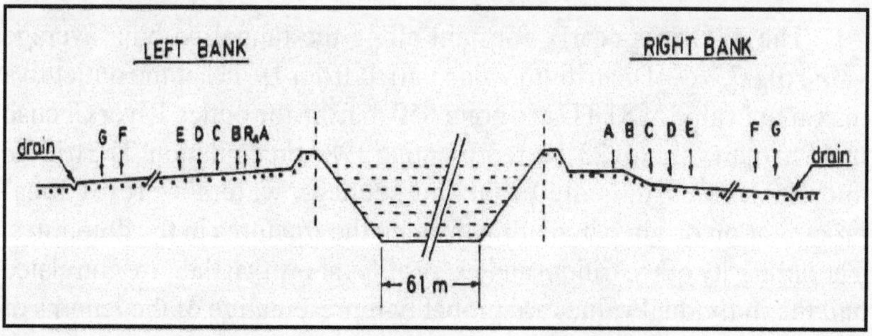

Figure 3.23 Ganga Canal cross-section (A, B, C— represent the boreholes on either side of the canal)

3.6.3 Seepage Source Identification in Tunnel

During the construction of a tailrace tunnel through a dolomite hill for the Salal Hydroelectric Project near Jammu, India several seepages of varying intensity were encountered. The tunnel, 2.4 km long and 11 m in diameter was intended to carry the tail waters and put them back into River Chenab, which hugs the tunnel. Tunneling was mainly through fractured rock (dolomites) highly jointed and often crumbly and sheared. High seepages were encountered whenever a shear zone was pierced. A huge seepage (~ 100 L/s) was encountered at about 800 m from the outlet where many shear zones intersect and the work had to be suspended for investigation of the source of seepage to ensure worker safety. Besides the River Chenab itself, a stream and a spring in the hillock were suspected to be possible sources of seepage.

A set of samples of seepage waters and of suspected sources was collected in July 1984 and analyzed for stable isotopes (^{2}H or D and ^{18}O), bomb tritium and electrical conductivity [21]. The δ D - $\delta^{18}O$ relationship of the samples is given in Fig. 3.24.

It can be seen from the figure the isotope composition of the seepage waters was distinctly different from those of the river and other surface waters in the hillock. The heavier isotope composition of River Chenab is highly depleted as it originates in the higher altitudes in the Himalayas. It is thus clear that the suspected surface waters were not contributing to the seepages.

The variation of $\delta^{18}O$, tritium (3_H) and electrical conductivity along the length of the tunnel is given in Fig. 3.25.

The $\delta^{18}O$ was nearly constant along the tunnel with an average value of –7.7‰. The tritium value varied from 19 TU at the outlet to a maximum value of 81 TU at a point 850 m from the outlet. River Chenab had a tritium value of 23 TU representing 1984 precipitation. The tritium and EC values indicated that the seepage waters were of local precipitation origin percolating through the fractures in the dolomites. The similarity of the tritium and EC profiles show that they are correlated and the individual values are probably representative of the lengths of the percolation paths along the shear zones.

The residence times of seepage waters could be estimated from the IAEA/WMO tritium-in-precipitation data of New Delhi and Kabul (Kazimir) stations [24] and assuming a piston flow model of percolation.

The table above shows that the seepage waters were 10 to 15 years old precipitation waters stored or percolating in the fractured dolomite. **The safety of the tunnel from seepage from any major source was thus assured**.

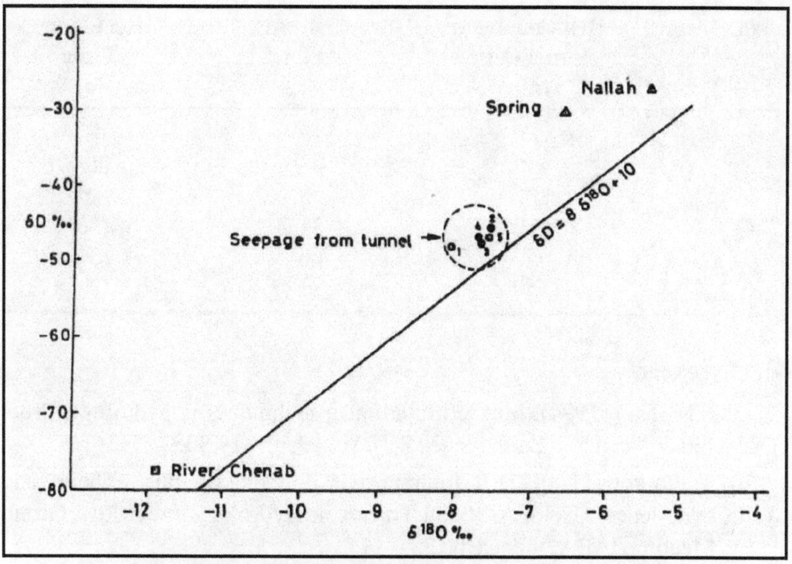

Figure 3.24 δ D - δ^{18}O plot of Salal tunnel seepage waters

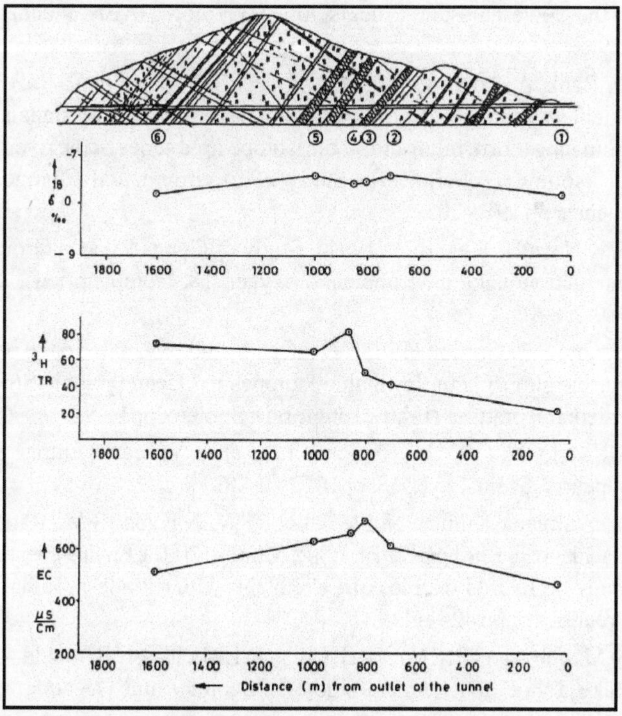

Figure 3.25 δ^{18}O, ^{3}H and electrical conductivity along the length of the tunnel

No.	Distance from Tunnel Outlet (m)	Tritium (TU)	Residence Time (a)
1	0	19	-
2	650	43	11.0
3	800	48	12.0
4	850	81	15.0
5	950	66	13.0
6	1550	72	13.5

References

1. S.M. Rao (1984), Injected radiotracer techniques in hydrology, Proc. Indian Acad.Sci (Earth Planet. Sci), Vol.93, pp 319-335
2. T. Florkowski (1991), Tritium in river flow rate gauging – Theory and experience, Use of Artificial Tracers in Hydrology, Proc. Adv. Group Meeting, IAEA, Vienna, pp 35 – 44
3. W.G. Mook (1970), Stable carbon and oxygen isotopes of natural waters in the Netherlands, In: Proc. Isotope Hydrology, IAEA, Vienna, pp 163-189
4. D. Rank, A. Adler, L. Arguás- Arguás, K. Froehlich, K. Razonski, W. Stichler (1998), Hydrological parameters and climatic signals derived from long term tritium and stable isotope time series of the river Danube, In: Isotope Techniques in Study of Environmental Change, IAEA, Vienna, pp 191 – 205
5. S.V. Navada, S.M. Rao (1991), Study of Ganga River – Groundwater interaction using environmental oxygen-18, Isotopenpraxis, 27, 8, pp 380 – 384
6. H. Craig, L. Gordon (1965), Deuterium and oxygn-18 in the ocean and marine atmosphere, In: Stable Isotopes in Oceanographic Studies and Paleotemperatures (Ed. E. Tongiorgi), Spoleto, pp1 – 22
7. M. Majoube (1971), Fractionnement en oxygéne-18 entre l'eau et la vapeur, J. Chim. Phys., 68, pp1423 – 1436
8. U. Saravana Kumar, Noble Jacob, S.V. Navada, S.M. Rao, Rm. P. Nachiappan, Bhishm Kumar, J.S.R. Murthy (2001), Environmental isotope study on hydrodynamics of Lake Naini, Uttar Pradesh, India, Hydrol. Process, 15, pp 425 – 439
9. D.M. Imbolden, R.F. Weiss, H. Craig, R.L Michael, C.R. Goldman (1977), Lake Tahoe geochemical study, Limnology and Oceanography, 22, pp1039 – 1051

10. Statistical Treatment of Data on Environmental Isotopes in Precipitation, IAEA Vienna, 1992, 781 pp (STI/DOC/10/331)

11. J.V. Turner, G.B. Allison, J.W. Holmes (1984), The water balance of a small lake using stable isotopes and tritium, Journal of Hydrology, 70, pp 199-220

12. P.R. Bierman, A. Albrecht, M.H. Bothner, E.T. Brown, T.D. Bullen, L.B. Gray, L. Turpin (1998), Erosion, weathering and Sedimentation, In : Isotope Tracers in Catchment Hydrology (Ed : C. Kendall, J.J. McDonnell), Elsevier, pp. 647-678.

13. U. Saravana Kumar, S.V. Navada, S.M. Rao, Rm. P. Nachiappan, Bhishm Kumar, T.M. Krishnamoorthy, S.K. Jha, V.K. Shukla (1999), Determination of recent sedimentation rates and patern in Lake Naini, India by ^{210}Pb and ^{137}Cs dating techniques, Applied Radiation and Isotopes, 51, pp. 97-105

14. A. Radhakrishnan Nair, A.S. Pendharkar, S.V. Navada, S.M. Rao (1979), Groundwater recharge studies in Maharashtra, Development of isotope techniques and field experience, In: Isotope Hydrology, IAEA, Vienna, pp803 – 826

15. A. Owczarczyk, S. Spilowski (1983), Sorption method for leakage localization in water reservoirs of Hydrotechnical plants, Isotope Hydrology 1983, Proc. of a Symposium, IAEA, Vienna, pp 709-717

16. V.K. Iya, K. Krishnamurthy, K.S. Aggarwal, R. Rajagopalan (1967), A Radiotracer Study on Flow path of Seepage in the Coffer Dams at Srisailam, Int. Jour. Appl. Rad. Isotopes, Vol.18, pp349 – 357

17. S.M. Rao (1975), Use of injected radioactive tracers in groundwater hydrology, Proc. Indo-German Workshop on Approaches and Methodologies for Development of Groundwater Resources, NGRI, Hyderabad, pp323-341

18. U.P. Kulkarni,—(1992), Radiotracer experiment to identify the location of seepage areas in the Borda minor irrigation tank, Yeotmal district, Maharashtra, Proc. Nuclear and Radiochemistry Symposium, Visakhapatnam, pp. 311-313

19. J. Roldão, N. Lopes, W. Stichler, A. Zuber (1989), Artificial and environmental tracers in the study of leakage under a dike of the Samuel reservoir, Porto Velho, Brazil, In: Isotope Hydrology Investigations in Latin Ameica, IAEA, Vienna, TECDOC-502, pp103 – 111

20. K.M. Kulkarni, K. Shivanna, Suman Sharma, A.R. Nair, U.P. Kulkarni, U.K. Sinha, S.V. Navada (1998), Environmental Isotope Study for a

Dewatering Operation at Jhamarkotra Mine, Rajasthan, Applications of Radioisotopes and Radiation in Industrial Development, Int. Conf. ICARID, Mumbai, pp. 350 - 354

21. S.M. Rao, S.K. Jain, S.V. Navada, A.R. Nair, K. Shivanna (1987), Isotopic study on sea water intrusion and interrelations between water bodies: Some field examples, In: Isotope Techniques in Water Resources Development, IAEA, Vienna, pp 403 – 421
22. K. Krishnamurthy, S.M. Rao, (1969), Theory and experiment in canal seepage estimation using radioisotopes, Journal of Hydrology, 9, pp 277 - 293
23. V.I. Aravin, S.N. Numerov (1965), Theory of fluid flow in undeformable porous media, Israel Programme of Translations, Jerualem, pp 191 - 193
24. IAEA, (1981), Statistical Treatment of Environmental Isotope Data in Precipitation, Technical reports Series No.206, IAEA, Vienna

Bibliography

1. Guidebook on Nuclear Techniques in Hydrology, (1983), Tech. Rep. Series No.91, IAEA, Vienna
2. Environmental isotopes in the hydrological cycle, Principles and applications (Ed. W.G. Mook), Introduction, IHP-V, Technical Documents in Hydrology, No.39, Vol.II (Atmospheric Water) & III (Surface Water), Unesco, Paris, 2000
3. Use of isotope techniques in lake dynamics investigations (2001), AEA-TECDOC-1206, International Atomic Energy Agency, Vienna

4

ISOTOPE APPLICATIONS IN GROUNDWATER HYDROLOGY

After precipitation, run-off and surface water storage in lakes and man-made reservoirs, the next steps in the hydrological cycle are infiltration through the unsaturated zone and groundwater dynamics in the saturated zone. Well-established applications of isotopes in groundwater hydrology are

1. Estimation of rate of soil moisture transport in the unsaturated zone using bomb tritium or injected radioactive tracers
2. Measurement of groundwater velocities using injected tracers in highly localized spots or using environmental tritium or carbon-14 for regional scale data.
3. Identification of existence of groundwater recharge and understand the recharge process (es)
4. Studying groundwater salinisation processes including seawater intrusion
5. Age-dating of old ground waters
6. Identification of palaeowaters in arid zones
7. Studying groundwater pollution from natural causes and human activities

4.1 Unsaturated Zone - Rate of Soil Moisture Transport

4.1.1 Use of Bomb Tritium

The piston flow model of soil moisture transport discussed in 2.2.1.1 was successfully applied for the application of the 1963 hydrogen bomb tritium peak in soil moisture to estimate rainfall recharge to groundwater [1].

The procedure involves collection of soil cores up to sufficient depth, extraction of soil moisture from the cores by vacuum distillation and measuring the tritium content at each depth and building the tritium profile. Interpretation involves identification of the 1963 peak position. The total soil moisture content above the peak is the rainfall recharge from 1963 to the date of investigation. An example of the method is shown in Fig. 4.1 [2].

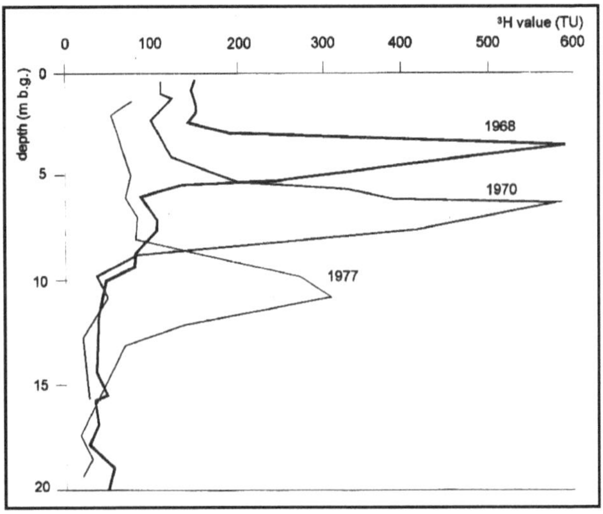

Figure 4.1 Example of downward movement of the 1963 tritium peak in unsaturated chalk in England

The technique was applied in several parts of Gujarat, India [3] and it was estimated that only a small fraction (5-10%) of rainfall went to recharge the groundwater.

An alternative approach is to determine the total tritium in the soil column up to the 1963 peak and dividing it by the total tritium in precipitation, i.e.

$$\textbf{Recharge} = (\Sigma H_i\ m_i) / (\Sigma H_y\ P_y)$$

Where H_i and m_i are the tritium content (TU) and moisture content (m) in each core i and H_y and P_y are the tritium content (TU) and precipitation (m) in each year y.

The environmental method is no longer applicable in most areas except probably in arid zones as the 1963 peak would have reached the saturated zone in all other places. In view of this an alternative injected tracer method [4] came to be used more extensively.

4.1.2 Injected Tracer Method

In 2.2.1.2 we discussed the basis of application of the injected radioactive tracer method using either tritiated water or $K_3\,^{60}C_O\,(CN)_6$ for assessing the rate of soil moisture transport in the unsaturated zone.

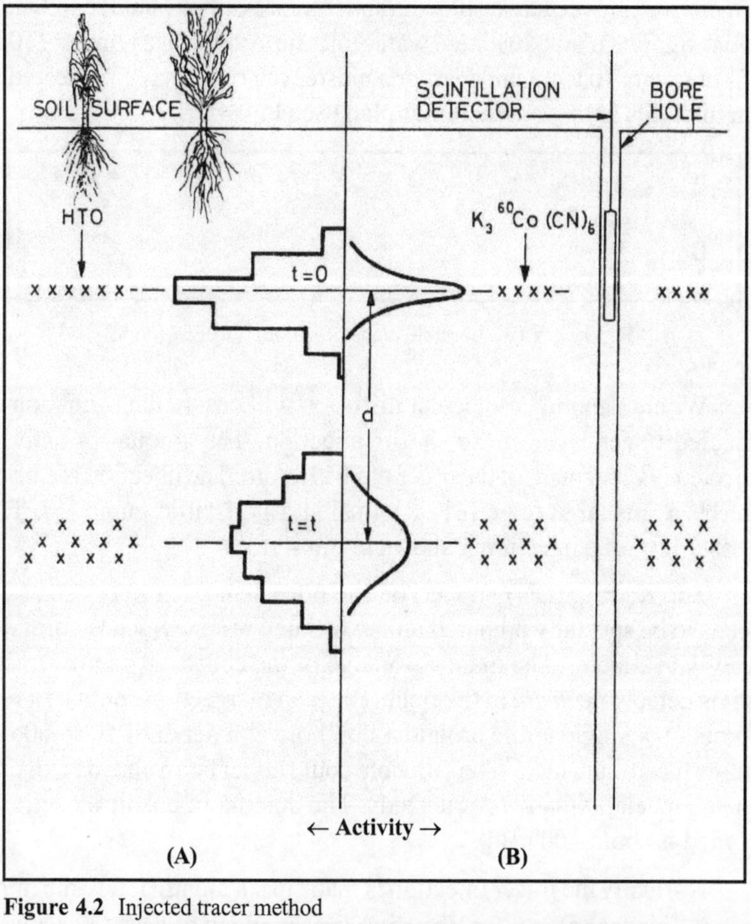

Figure 4.2 Injected tracer method
(A): Tritium Tracer Method, Detection by liquid scintillation technique
(B): Gamma Tracer Method, Detection by NaI scintillation detector

The tracer method is as shown in Fig. 4.2. The soil moisture below the root zone (50 to 80 cm below ground level) is tagged with the selected tracer, HTO or $K_3\,{}^{60}C_O\,(CN)_6$. Injection below the root zone is to avoid soil moisture uptake by the plants and consequent transpiration.

Tracer injection

Tritium tracer is most commonly used as it is closest to water in all respects and is also very convenient from the point of radiation safety. The disadvantage is the relatively more cumbersome assay procedure of collecting soil cores and the need for laboratory facilities to extract soil moisture by vacuum distillation and tritium assay by liquid scintillation counting. The plan of tritiated water injection could be a number (10 to 15) of points 10 to 15 cm apart or a cluster of points (say 5) repeated at 1m intervals or any other similar plan (See Fig. 4.3).

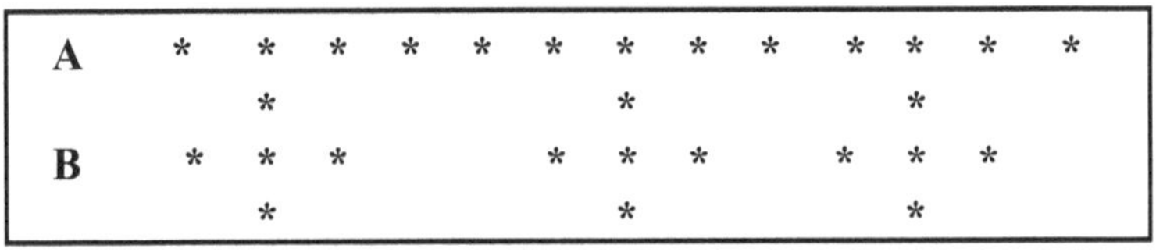

Figure 4.3 Two possible plans of tritium tracer injection

Within a month, molecular diffusion will ensure the formation of injected tracer layer at the site of injection. The amount of activity injected is very small, of the order of 50 kBq/dm. The injection is carried out by a special syringe [4] or by using a peristaltic pump [5]. The peristaltic pump injection is shown in Fig. 4.4.

The $K_3\,{}^{60}C_O\,(CN)_6$ tracer, on the other hand, is rarely used as it needs to be specially prepared unlike tritiated water, which is normally readily available. The main advantage of the cobalt-60 gamma tracer is it is detectable *insitu* in the field. The plan of injection could be a few points on a single circle around a borehole at a depth of 50 to 80 cm below the ground level. The borehole could either be an already existing one or specially made for the study. The amount of cobalt-60 activity needed is about 400 kBq.

Normally the tracer injection is made just a month before the rainy season (monsoon) and the detection is carried out at least once a year thereafter.

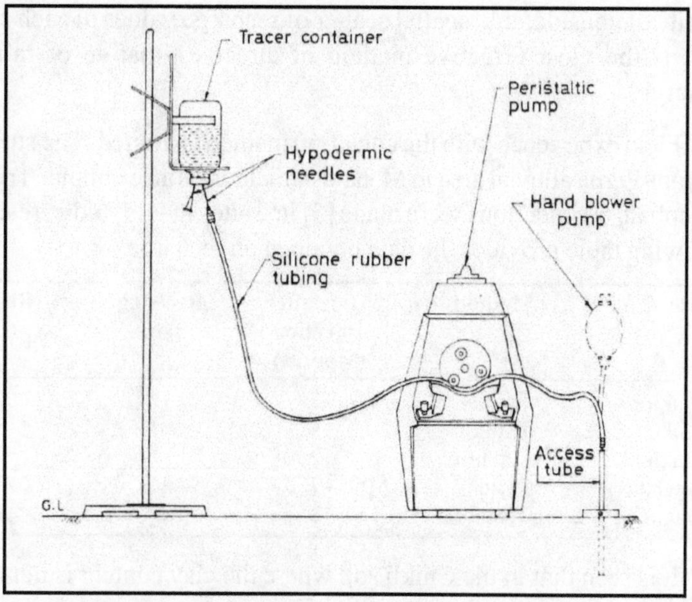

Figure 4.4. Tracer injection by peristaltic pump

Field examples

Since early seventies, a large number of recharge measurements were made in India using the injected tritium method. The early studies were carried out mainly in Northern India, particularly in the Indo-gangetic plain [6]. In measurements carried out in about 80 sites, the recharge values varied from 6 to 20 cm amounting to 12 to 20% of precipitation.

A number of measurements made in the Sabarmati basin of Gujarat [7] during the two-year period of 1967-68 showed that the average recharge to the unconfined aquifers was of the order of 8% of precipitation and the spatial variability was governed by the silt and clay content (<45ìm) of the soil.

In a study [8] comprising of 26 tritium injections in Lower Maner basin (1600 km^2) in Southern India, the recharge varied from 5 to 24 cm (Mean: 10 cm) for a precipitation of 125 cm. This compared favourably with water table fluctuations.

Compilation [9] of data collected over 25 years from tritium tracer experiments in many parts of India indicated that the tracer method is, in spite of considerable spatial scatter of recharge values in each study area, is the most effective method of direct estimation of rainfall recharge.

Field experience with the cobalt-60 method is limited. The study in the Tapi-Purna alluvial area in Maharashtra, India is an example. Tritium and cobalt-60 injections were made [5] in 2 sites in this study area. The following table provides the data obtained on recharge.

Site	Method	Time after Injection (Months)	Recharge (cm)	Rainfall (cm)
Chikli	Tritium	12	8	72
Chikli	Tritium	18	8	88
Chikli	Cobalt-60	12	5	72
Waghoda	Tritium	12	12	79
Waghoda	Cobalt-60	12	19	79

It is seen that in the Chikli soil where the clay content is not high, the two tracers (not injected at the same spot) are more comparable (Fig. 4.4) whereas in the Waghoda soil where the clay content is large, the trivalent $[Co(CN)_6]^{3-}$ ion appears to move ahead of tritium due to anion exclusion [10] as the clay particles are negatively charged.

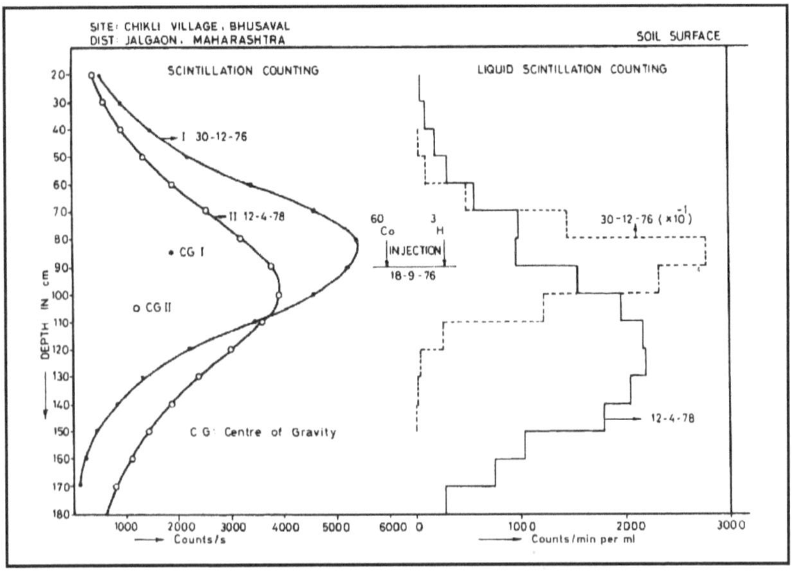

Figure 4. 5 Comparison of $^1H\,^3HO$ and $K_3\,^{60}C_O\,(CN)_6$ in a field study

4.2 Saturated Zone

Isotope hydrology of the saturated zone includes determination of groundwater velocity, identification of recharge and the characterization of the recharge processes, salinisation processes including seawater intrusion and groundwater pollution from both natural and anthropogenic sources. These aspects are covered in this section.

4.2.1 Groundwater Velocity

4.2.1.1 Use of injected tracers

Measurement of groundwater velocity either in a single borehole or at a given site or in an aquifer system is of considerable importance to assess aquifer parameters leading to resource estimates or to predict pollutant progress in specific cases (see Eq. 2.14).

In 2.2.13, tracer methods for the determination of groundwater velocities, both filtration (Darcy) and interstitial (true), have been discussed. In 3.6.2, an example of application of the point dilution method for determination of permeability of canal seepage medium is given.

Field examples of multi-well method

Venice plain, Italy

In a multiwell tracer study in the alluvial plains of Venice, Italy radioactive ^{131}I was successfully used [11] to determine the interstitial groundwater velocity and aquifer dispersivity over a distance of 25 m. The objective was to assess the hydrodynamic properties of the phreatic aquifer from the point of view of design of pollution control measures. The experimental site is shown in Fig. 4.6.

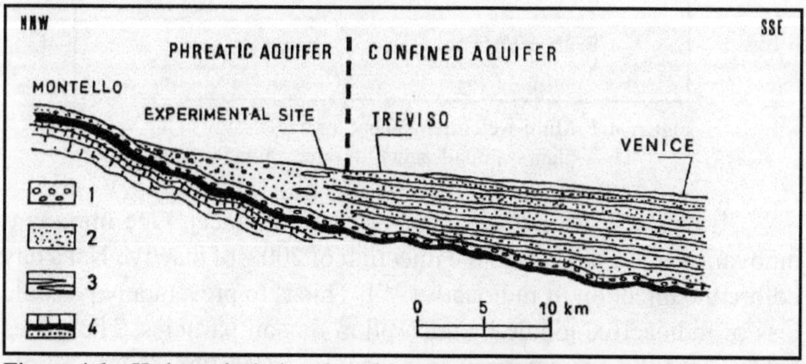

Figure 4.6 Hydrogeological section of the Venetian alluvial plain. 1: coarse sediment, 2: fine sediment, 3: clayey layers, 4: pre-quaternary substratum

The groundwater table was originally 23 m below the ground level. The experimental site was a disused quarry and hence the water table was only 2-3 m below the ground surface. Fig. 4.7 shows the plan of the 'Tracer injection' well and other detection wells downstream.

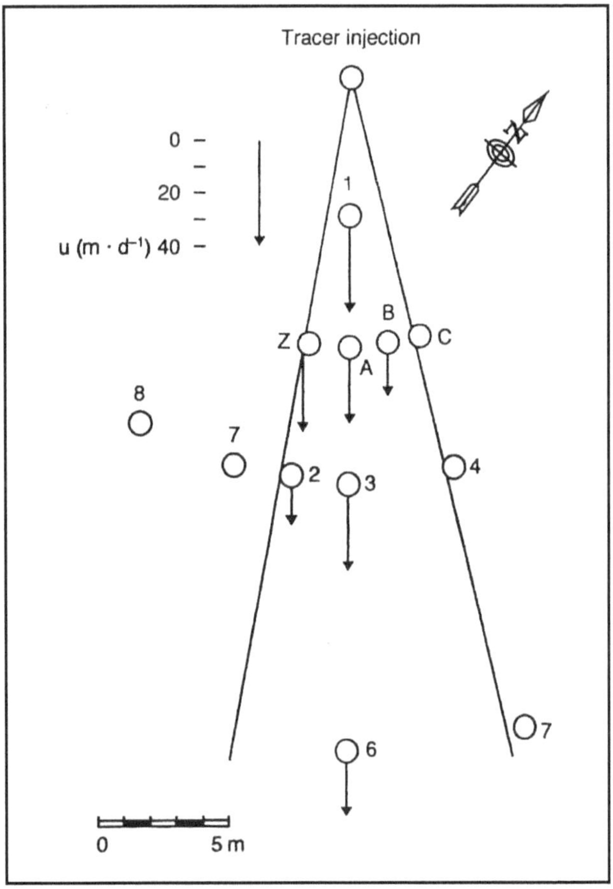

Figure 4.7 Multi-well arrangement for tracer study in the Venice plain - Groundwater flow direction and velocity vectors

^{131}I of about 1 GBq was used as the radiotracer. One important innovation in the study was the injection of 200 g of inactive NaI a day before the injection of radioactive ^{131}I. This is to prevent any possible loss of radioactive iodide by adsorption on soil particles. Though as halide iodide is not supposed to be lost by adsorption, there is some evidence [12] that it might undergo some biogeochemical reactions converting it into some insoluble form. By injecting substantial quantity of inactive iodide, the probability of active iodide undergoing the undesirable reaction is minimized.

After injecting the radiotracer, radioactivity was measured simultaneously in the boreholes at different depths. The tracer logs made it possible to localize the different pathways and to calculate the effective velocity, which varied from 11 to 33 m/day.

Using the dispersion equation similar to Eq. 2.14, the longitudinal dispersion coefficient D_L was calculated to be about 86 m^2 /day. Recalling the longitudinal dispersivity $a_L = D_L/v$ where v is the flow velocity, the dispersivity ranges from 2 to 5 m.

Manikaran Geothermal Field

Manikaran is located in Himachal Pradesh, India and on the bank of Parbati River, a tributary to Beas. The geothermal field at Manikaran was being investigated for its potential for power generation. As part of this investigation, the direction of movement of the deep geothermal fluids needed to be determined.

One opinion was that the thermal waters rose along the Kulu thrust plane north of the Parbati River (Fig. 4.8) and the deep waters flowed southward through the highly jointed and fractured quartzites to emerge as hot springs on both the banks of the river. The second opinion is that the deep geothermal waters flowed along the Parbati River i.e. east to west.

Several bore wells including two deep wells were made for the main investigation project and were used for the multi-well tracer experiment [13] to determine the direction and movement of the deep geothermal fluids. About 1 GBq of tritium as tritiated water was injected at a depth of 130 m in the Deep Geothermal Well DGW-1 (Fig. 4.8), north of Parbati River. Daily samples were collected from all the available bore wells for tritium analysis in the laboratory by liquid scintillation counting. The results showed that significant tritium levels (twice the background level or more) were detected only in three wells; 21, II and IV.

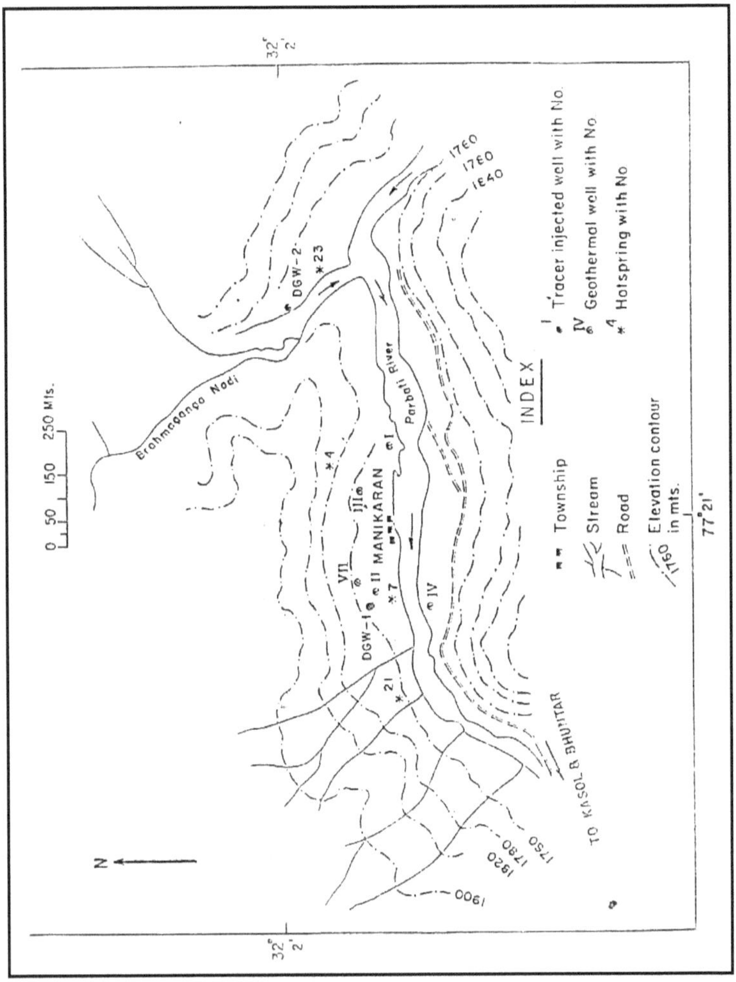

Figure 4.8 Manikaran experiment site

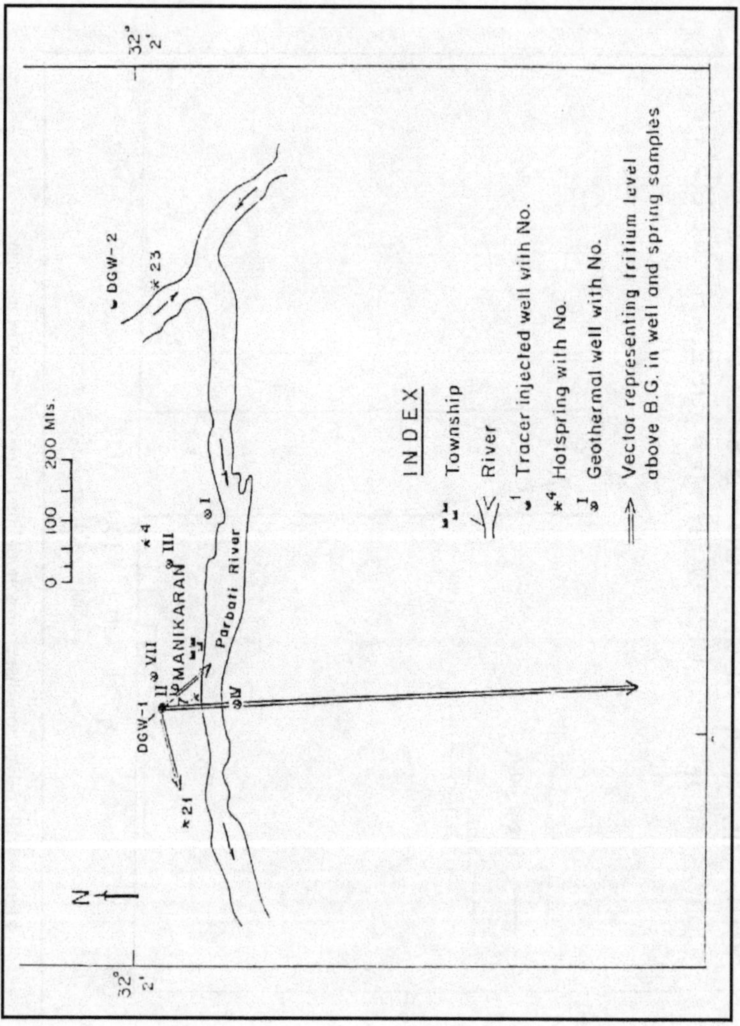

Figure 4.9 Relative maximum tritium levels detected in Manikaran wells

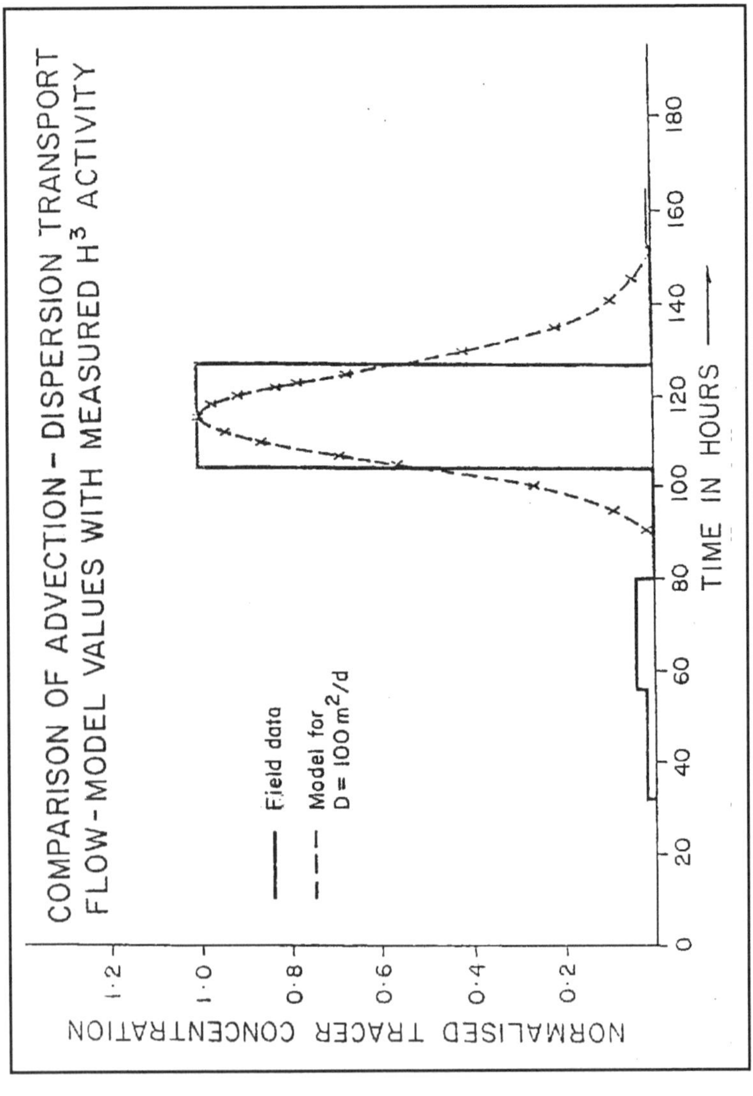

Figure 4.10. Advection-dispersion modeling of tracer data in Manikaran

Fig. 4.9 shows the relative maximum tritium levels detected in the above three wells. The maximum observed in well IV was the highest and was recorded 5 days after injection. It could thus be seen that the predominant direction of transport is ***southward across the Parbati River*** with a velocity of 70 m/day

The advection – dispersion model (Eq. 2.14) was applied to the Manikaran result to understand why only one single day's measurement was significant. Fig. 4.10 shows the best fit was obtained for a dispersion coefficient of 100 m^2/day. The sampling should have been more frequent than once in a day to obtain more than one significant measurement.

4.2.1.2 Groundwater Velocity through ^{14}C dating

The hydrologist is often required to age date slow moving deep groundwater in arid and semi-arid regions. These waters are normally old and are amenable for carbon-14 method of age determination as described in 2.4.1. Determination of groundwater age at different points along the flow path can yield information on the velocity of flow. Irrespective of the type of correction made for carbon-14 age (see 2.4.1), the velocity value remains undisturbed as the time scales in different models appear to be linearly related.

Some examples of groundwater velocity measurement through carbon-14 age determination are given in subsequent sections. (See 4.2.3.1, 4.2.3.3)

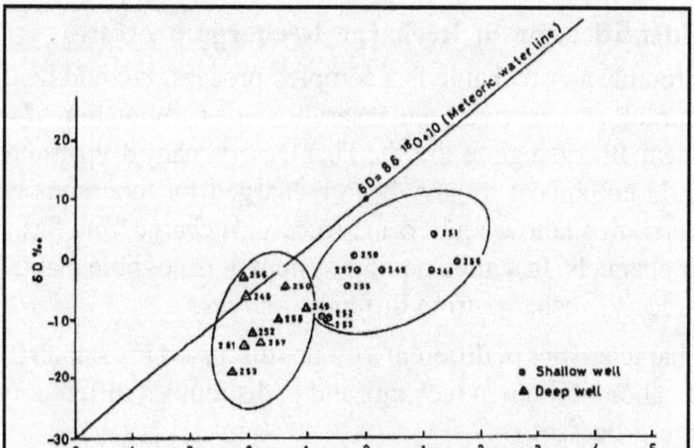

Figure 4.11 δD - $\delta^{18}O$ plot of Cauvery delta groundwater samples

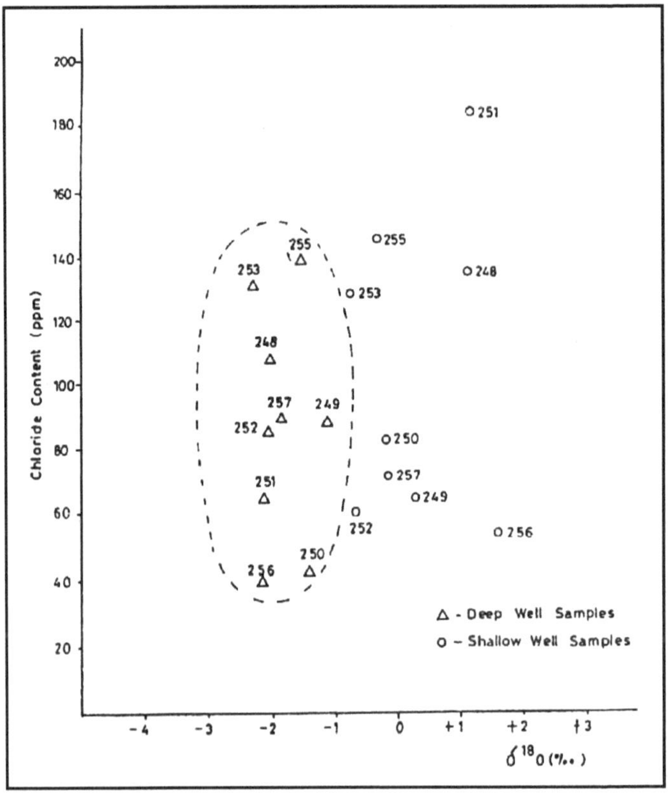

Figure 4.12 $\delta^{18}O$ – Cl^- plot of Cauvery delta groundwater samples

4.2.2 Identification of Recharge Recharge processes

Groundwater recharge is a complex process. It could be directly from rainfall or from rivers and reservoirs or a combination of some or all of them. In arid regions like the Thar Desert, many deep groundwater bodies do not appear to have been recharged for thousands of years whereas some shallow aquifers may receive recharge from flash floods in river channels. In a multi aquifer system, it is possible that different layers receive recharge from different sources.

Characteristics of different isotopes discussed in Chapter 2 enable identification of modern recharge and to distinguish different types of recharge processes.

A couple of examples pertaining to groundwater recharge were mentioned in Chapter 3. In 3.4, the use of a stable isotope approach for the identification of groundwater recharge from a percolation tank was discussed. In 3.5.2, it was seen how sable isotope composition ruled out the possibility of reservoir contribution to groundwater in Jhamrakotra phosphate mine. The following examples will further highlight the role of environmental isotopes in groundwater recharge investigations.

4.2.2.1 Shallow and Deep Aquifers in Cauvery Delta

Cauvery delta in Tanjavur district is the rice bowl of Tamil Nadu, South India and often faces severe shortage of irrigation water. A hydrogeological study was underway in the eighties to look at the possibility of artificial recharge of the deltaic ground waters. The delta consists of Quaternary alluvium consisting of sand, silt and clay to a depth of 30 m followed by Pliocene formation. The shallow aquifer in the Quaternary was at a depth of 10 to 20 m whereas in the Pliocene formation it was at depths ranging from 40 to 80 m. Geologically, the deposition in the shallow and deep aquifers is very erratic and the thickness of the intervening clay, clayey sand and gravel also varies.

From the geology and the piezometric levels, it could not be confirmed how well the two aquifers were interconnected, a crucial piece of information to plan artificial recharge measures.

Water samples from piezometers in nine different locations were analyzed [14] for δD, $\delta^{18}O$ and chloride. From the δD - $\delta^{18}O$ and $\delta^{18}O$ – Cl^- diagrams (see Figs. 4.11, 4.12), it can be clearly seen that the shallow and deep aquifer waters fall into two distinctly different groups. The shallow samples generally show enriched stable isotope values. Isotope enrichment could be possibly due to recharge to the shallow aquifer from the local surface channels whereas the source of recharge to the deeper aquifer could be from a remote outcrop area. The two groups meet at places where the intervening clayey layers are very thin. **In short, the two aquifers are generally not interconnected.**

4.2.2.2 Effect of Storm on Groundwater Recharge

Bagepalli area (935 km^2) located in the northern part of Kolar district in Karnataka state in South India is semi-arid with an annual rainfall of 550 mm. The area is made up of granites and gneisses and belongs to the peninsular gneissic complex. Groundwater occurs in lineaments, weathered valley portions and in open joints and fractures.

The soil cover is generally 1-2 m, but extends up to 20 m in some places.

An isotope study on the hydrometeorology and groundwater recharge [15] in the area coincided with a storm event (600 mm) of November 1991.

Surface water and groundwater samples were collected in October 1991 (pre-storm) and January 1992 (post storm) and analyzed for ^{2}H, ^{18}O and tritium. A sample of the rainfall was also collected for analyses.

$\delta^{2}H$ of pre-storm groundwater:	-10 to -23‰ (Average: 22‰)
$\delta^{18}O$ of pre-storm groundwater:	-1.7 to –3.6‰ (Average: 3.1‰)
$\delta^{2}H$ of post-storm groundwater:	-13 to -44‰ (Average: 28‰)
$\delta^{18}O$ of post-storm groundwater:	-2.1 to –7‰ (Average: 4.4‰)
$\delta^{2}H$ of storm water:	-65‰
$\delta^{18}O$ of storm water:	-8.9‰
Tritium in pre-storm groundwater:	3 to 7 TU
Tritium in post-storm groundwater:	4 to 8 TU

The depleted stable isotopic composition and higher tritium levels in post-storm groundwater show recharge. It could be calculated, based on isotopic balance (Eq.III.3) that the recharge component of the storm was about 19 to 27%.

4.2.2.3 Groundwater dynamics in Budapest aquifer

Drinking water (0.7x10^{6} m^{3}/d) to nearly 2 million population of Budapest, Hungary comes from bank filter wells (10 to 20 m deep) of the riparian aquifer system of River Danube. Majority of the wells are located in Szentendre Island to the north of the city and Csepel Island to the south (Fig. 4.13). It was necessary to study [16] the groundwater dynamics, as the system is vulnerable to pollution. Municipal wastewater is let out into the Danube River south of the city. This can have an effect on the groundwater in Csepel Island. In some areas the shallow ground water is subject to contamination from agricultural, industrial and communal activities. The Quaternary deposits in both islands are Pleistocene gravel and sand as well as Holocene gravel, sand, mud and clay. The groundwaters are generally Ca-Mg-HCO_3 type.

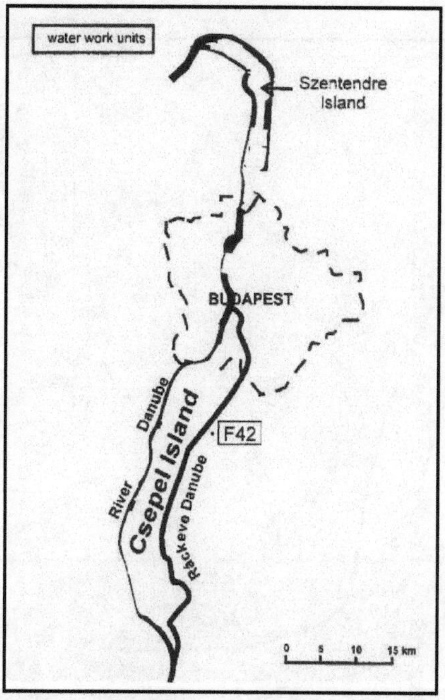

Figure 4.13 Map of the investigated area around Budapest

Isotope study

The Danube River water and the shallow groundwater recharged by infiltrating rainwater had distinctly different $\delta^{18}O$ values. Mean $\delta^{18}O$ of Danube was –11.0‰ (based on measurements made at Bratislava for several years) whereas the mean $\delta^{18}O$ of locally infiltrated phreatic groundwater was –9.3 ±0.4‰. Taking advantage of the difference it was possible to calculate the % contribution of Danube (X_{Danube}) to well water using the following equation, which is similar to Eq.3.3

$$X_{Danube} = \frac{\delta^{18}O_{well} - \delta^{18}O_{shallow\ groundwater}}{\delta^{18}O_{Dnube} - \delta^{18}O_{shallow\ groundwater}}$$

A large number of samples were collected at regular intervals from the Danube River and also from the production and observation wells and were analyzed for ^{18}O. Some samples were analyzed for Chloride, nitrate and sulphate to help identify the source of pollution in shallow groundwater.

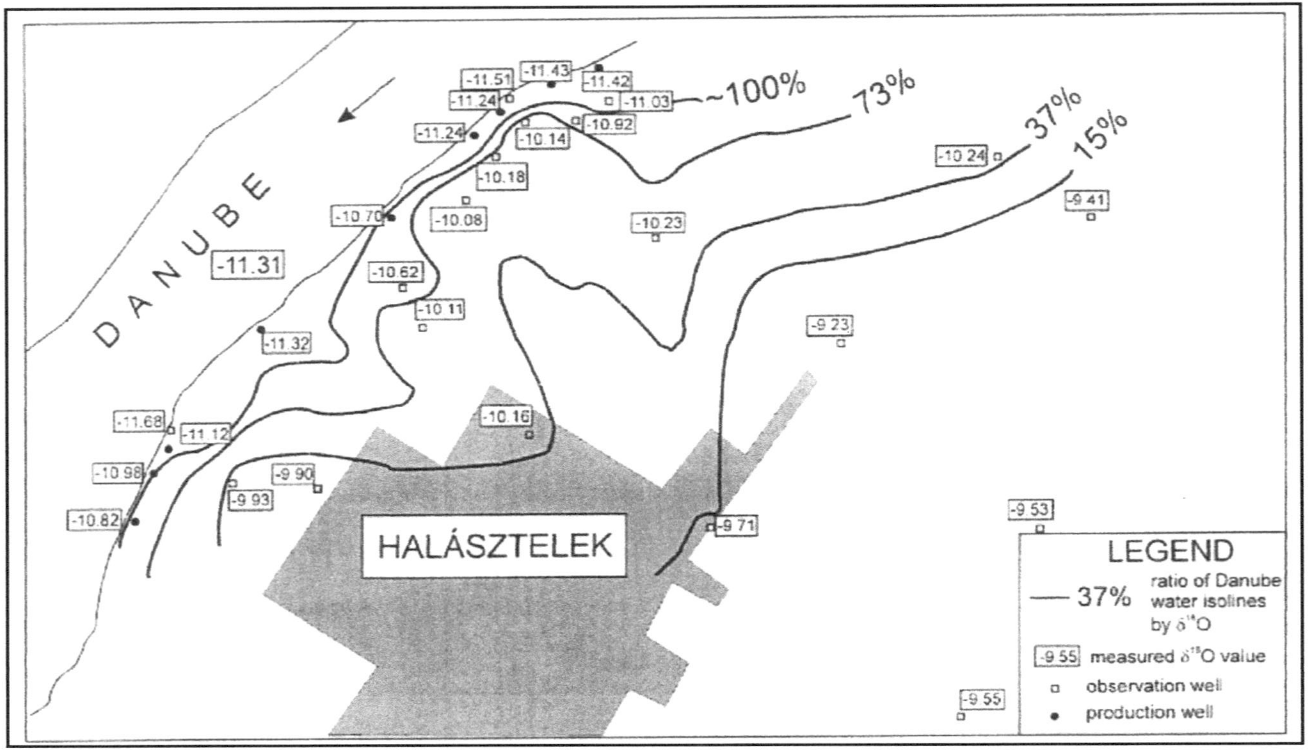

Figure 4.14 Contribution of Danube water at different distances from the river in the Halásztelek water works area

The above equation was applied to the area of Halásztelek water works (main pumping station) in Csepel Island and the resulting spatial distribution of infiltrated Danube waters is shown in Fig. 4.14. It could be seen that the production wells pumped out predominantly Danube water.

Seasonal variation of $\delta^{18}O$ of Danube and corresponding values for one Halásztelek production well H5 (close to Danube, not indicated in Fig. 4.14) are shown in Fig. 4.15. Snowmelt in the Alps in the spring and early summer results in most depleted $\delta^{18}O$ value in Danube waters. This negative peak in the $\delta^{18}O$ series could be used for calculating the transit time to the H5 well. A mathematical correlation programme of the two $\delta^{18}O$ series yielded a transit time of 24 days in summer when the Danube River level was high and exploitation was intense and 40 days in autumn/winter when the river level was low and exploitation was not so intense.

The SO_4, Cl and NO_4 levels correlated positively with $\delta^{18}O$ with the ion concentrations increasing with increasingly positive Oxygen-18. This clearly indicated that the source of pollution was not the Danube River, but the agricultural practices and municipal effluents.

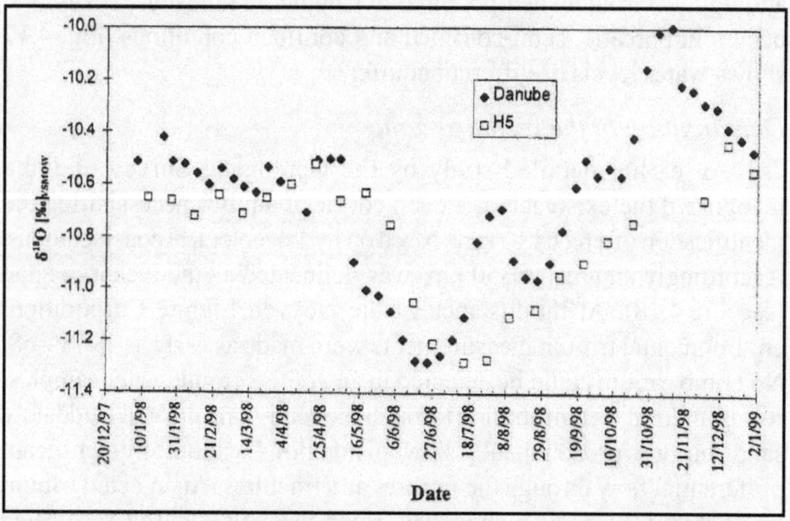

Figure 4.15 $\delta^{18}O$ time series of the Halásztelek (H5) production well and the smoothened $\delta^{18}O$ time series of Danube River

4.2.2.4 Estimation of Recharge to Neyveli confined aquifer

Neyveli area in coastal Tamil Nadu in South India has large lignite deposits. These deposits overlie the highly pressurized artesian Neyveli aquifer. This Cuddalore Sandstone aquifer had a hydrostatic pressure of 50 – 100 m above the lignite bed before the start of mining in 1961. Open cast mining of lignite necessitated depressurization of the aquifer with a withdrawal rate of 8000 to 9000 m^3/h. This rate of withdrawal coupled with the possibility of seawater intrusion needed an evaluation of the recharge to the confined aquifer.

An detailed isotope investigation [17] helped to first delineate the principal recharge area of the confined aquifer using environmental carbon-14 data followed by using estimates of direct rainfall recharge in the identified recharge zone made by injected tracer methods (see 2.2.1.2 and 4.1.2). Leakages from the phreatic and semi-confined aquifers to the confined aquifer were not considered significant

Fig. 4.16 shows the geology of the Neyveli study area. The Archaeans are exposed to the west, followed by Cretaceous, Tertiary and Recent alluvial deposits to the east. The Tertiary sediment deposits known as the Cuddalore formation occupy the major portion of the basin and consist of semi-consolidated sandstone, water-baring sands and lignite. Groundwater potential in Cuddalore sandstone is vast and occurs in phreatic, semi-confined and confined conditions. Fig. 4.17 shows water levels of different aquifers.

Demarcation of the recharge zone

An earlier detailed study by the Geological Survey of India recognized the existence of a deep confined aquifer necessitating the identification of a recharge zone based on hydrogeological considerations. Accordingly an area of 350 km^2 was delineated as the recharge zone (see Fig.4.18). At the instance of the Neyveli Lignite Corporation, environmental tritium measurements were made as early as 1966 – 68. No bomb tritium could be detected in any of the groundwater samples. It was inferred that infiltration through the semi-consolidated Cuddalore sandstone was predominantly slow 'piston flow' without any significant preferential flow through the fissures and fractures. An injected tritium tracer study [18] in mid-eighties confirmed piston flow and also estimated that the infiltration time would be the order of a few decades (see Fig.

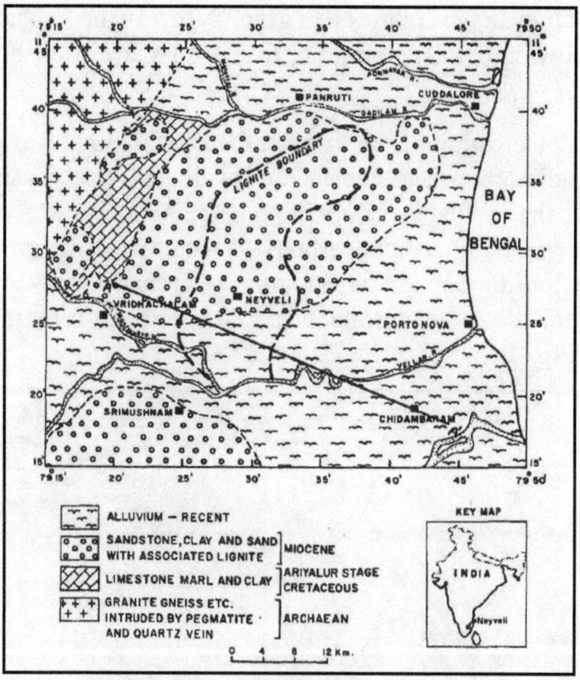

Figure 4.16 Geology of Neyveli study area

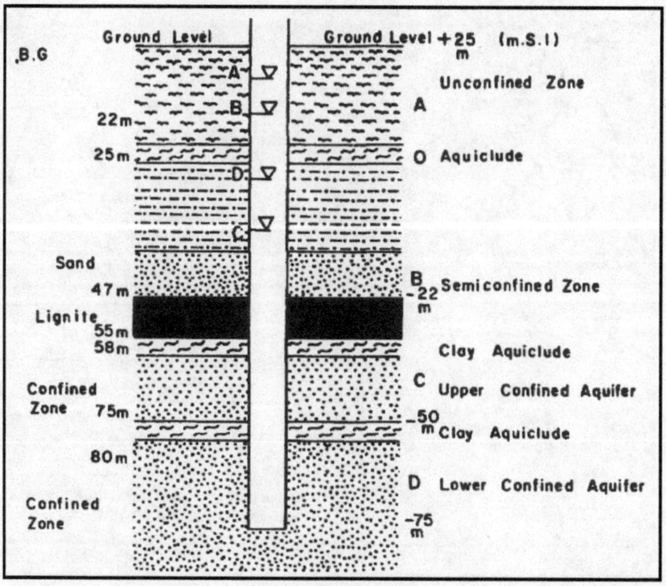

Figure 4.17 Cross-section of aquifers and their relative water levels

4.19). It is only in the eighties that groundwater in the 'recharge ' area started showing bomb tritium, i.e. nearly 15 years after the test ban treaty.

The demarcation of the Principal Recharge Zone was primarily based on radiocarbon measurements and hydrogeological considerations. Therefore, the groundwater in the recharge area was expected to be young (having bomb tritium/carbon-14) with increasing groundwater age in the flow direction. A large number of groundwater samples were analyzed for ^{14}C and the groundwater ages derived from the data are shown in Fig.4.15.

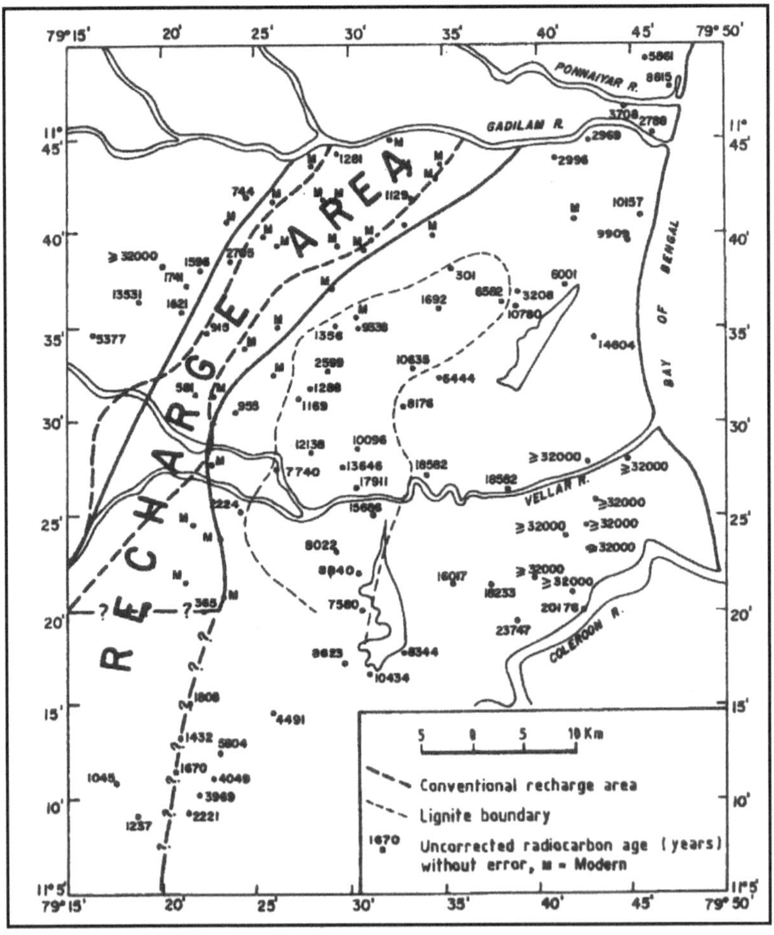

Figure 4.18 Groundwater ages leading to demarcation of the recharge area

The western boundary of the recharge area is governed by the Cretaceous – Tertiary contact and the eastern boundary by the combined confinement due to the associated clay beds sandwiching the lignite bed. In the extreme west uncorrected ^{14}C ages varied between ≥ 32000 and 2000 years with 'Modern' age samples towards east. Possible dilution of ^{14}C by Cretaceous limestone pushing up the apparent ^{14}C age was taken into consideration while drawing the recharge boundary. On the contact zone, except for a few with ages of 1500 – 2000 years, most of the samples showed 'Modern' age. Nearly 75% of the samples between the Cretaceous – Tertiary boundary and the western lignite boundary showed 'Modern' age. These samples were also largely within Vellar River in the southwest and the Gadilam River in the northeast. As the water leaves the recharge zone, the ^{14}C age increases in the flow direction. The recharge area thus demarcated worked out to be more than 600 km^2. This is almost twice the area that was earlier demarcated from hydrogeological considerations. This could further increase with time due to the continuous depressurization of the aquifer.

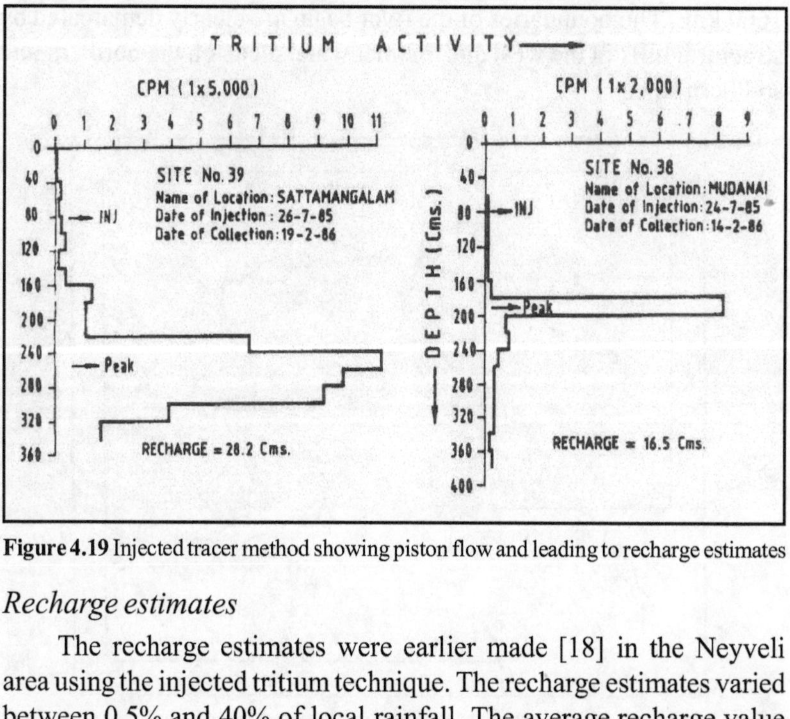

Figure 4.19 Injected tracer method showing piston flow and leading to recharge estimates

Recharge estimates

The recharge estimates were earlier made [18] in the Neyveli area using the injected tritium technique. The recharge estimates varied between 0.5% and 40% of local rainfall. The average recharge value for 34 sites within the demarcated recharge area worked out to be 15.5%.

Over a recharge area of 600 km^2, the total quantity of infiltrating water works out to be about $100x10^6$ m^3 /year.

This is probably a lower limit as the recharge area may extend further south from the Vellar River due to continued depressurization and also inter-aquifer transfer of groundwater was not considered. This evaluation will greatly help for groundwater management in the long term as the confined aquifer would continue to be pumped to meet the mining, agricultural, industrial and domestic needs.

4.2.2.5 Groundwater recharge to karstic aquifer in Amner River Basin [19]

Introduction

Amner River basin is a major part of the western Chhattisgarh basin, Rajnandgaon district in central India (Fig. 4.20) with an area of 1660 km^2. The boundaries of the river basin are clearly demarcated by structural hills in the west and distinct watersheds on the northern and southern sides.

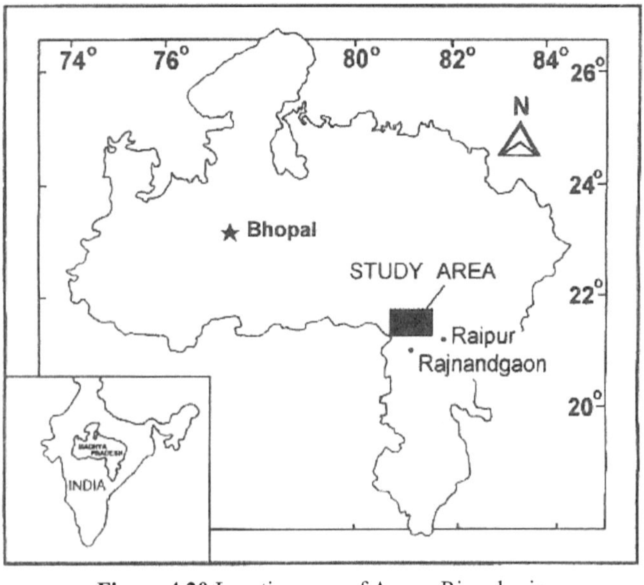

Figure 4.20 Location map of Amner River basin

In the study area Amner River forms the principal surface drainage. It flows west to east and joins the Shivnath River. The underlying rock formations; dendritic drainage in the limestone and sandstone in the plains and parallel drainage in the structural hills comprising metasedimentary and igneous rocks, mainly control the drainage. The area is of tropical sub humid type with an annual mean precipitation of 1050 mm.

A large reservoir and a spring are situated at higher elevations in the metasedimentary and igneous formations.

Development of secondary porosity in karst – limestone and dolomite with distinctive hydrogeology and landforms arising from a combination of high rock solubility and well-developed porosity is usually erratic and unpredictable. The isotope-chemical investigation was made to study the role of the karstic terrain in surface water – groundwater interconnections, groundwater recharge and karstic capture phenomenon.

A systematic well inventory revealed substantial fluctuation in water levels between the pre-monsoon and post-monsoon seasons. The groundwater occurs in phreatic conditions. The yield data analysis of 355 wells brought out that 82% of wells in limestone had poor yields and the distribution of successful wells in limestone was erratic. This was attributed to the anisotropic nature of limestone due to karstification.

Isotopes and chemistry

A large number of samples were drawn from various sources like precipitation, reservoir, river, spring, tube wells and dug wells and analyzed for major ions and isotopes. Sample locations are shown in Fig. 4.21.

Most of the groundwaters are fresh with EC less than 1100 μS/cm. EC ranges from 75 to 470 μS/cm for surface waters and 280 to 2300 μS/cm for groundwaters, which are of Ca-Mg-HCO_3 type common to limestone terrain.

Nitrate levels in surface waters range from 4 to 32 mg/L with an average value of 17 mg/L, 8 to 32 mg/L in tube wells with an average value of 17 mg/L and in dug wells the range is from 4 to 609 with an average value of 117 mg/L

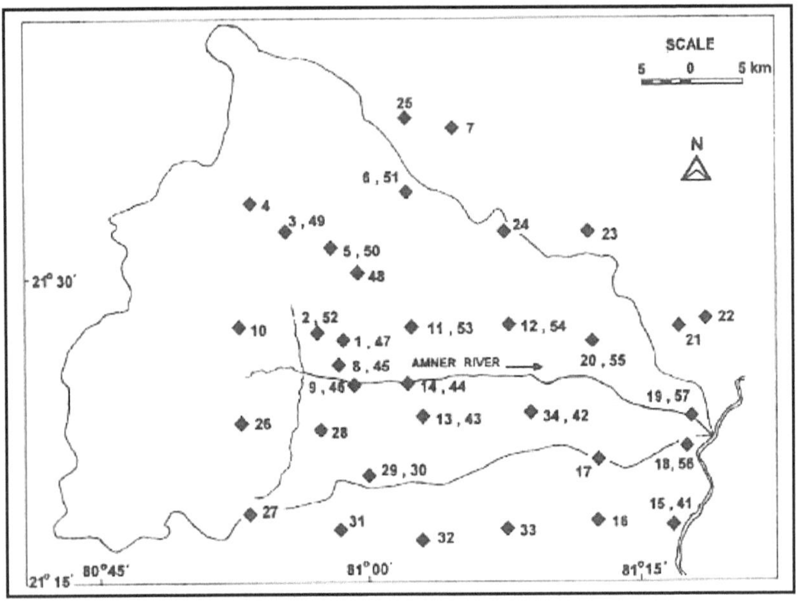

Figure 4.21 Sample location map of the Amner area

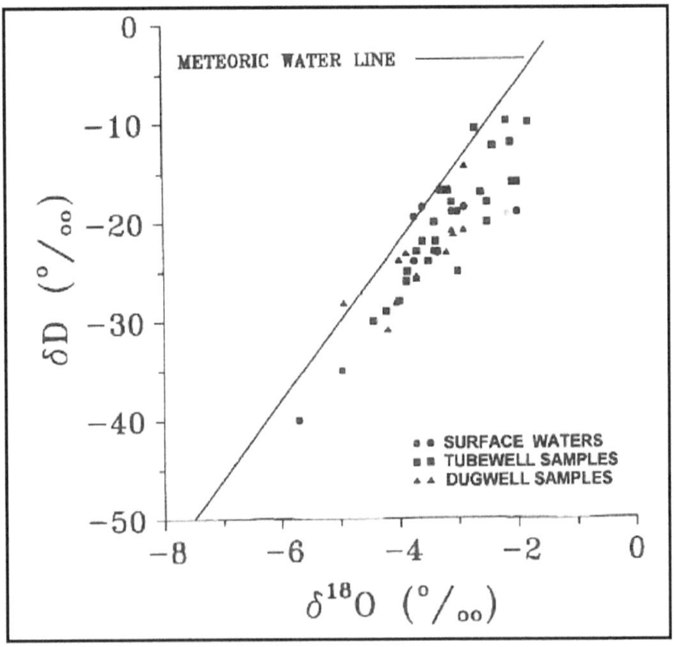

Figure 4.22 δD - δ^{18}O plot of the Amner basin samples

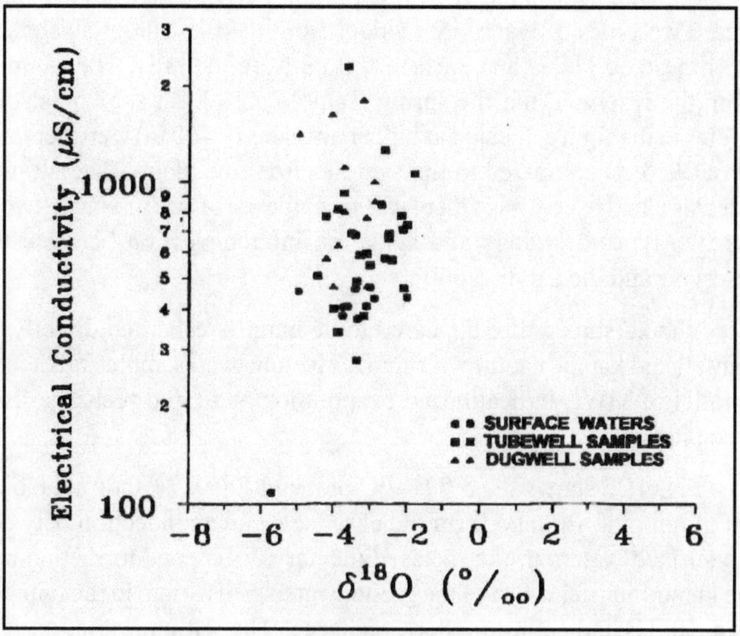

Figure 4.23 $\delta^{18}O$ – EC plot of the Amner samples

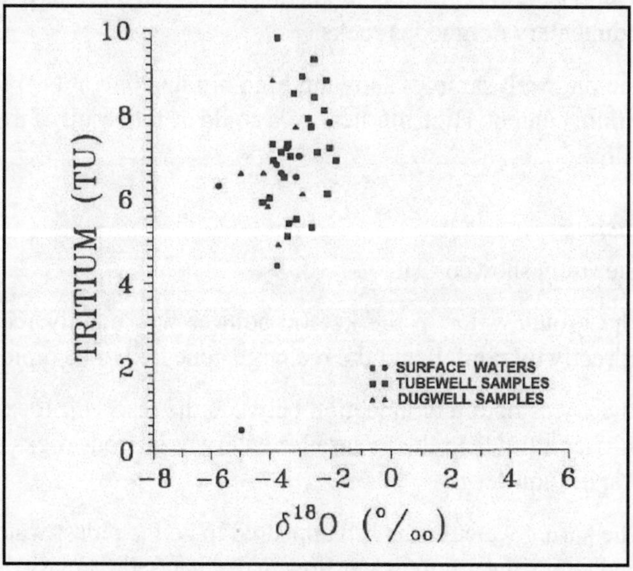

Figure 4.24 $\delta^{18}O$ – Tritium plot of Amner samples

Fig. 4.22 shows the δD - $\delta^{18}O$ plot of the samples from the study area. Two types of water are evident from the plot; groundwaters in the limestone plains and surface waters from the hills. The samples from the reservoir and the spring (sample nos. 3 &4 shown as solid circles in the figure) located at higher altitudes (~ 400 m) were depleted in δD & $\delta^{18}O$ compared to the samples from the plains. The altitude effect is clearly seen. Also there was no evidence of any mixing between the two types of waters and hence no interconnection between the reservoir and the karstic aquifer.

The karstic aquifer appeared to be mainly recharged directly by rainfall and karstic capture of runoff. The tube well samples fall nearly parallel to MWL indicating no evaporation and rapid recharge from precipitation.

Fig. 4.23 shows the $\delta^{18}O$ – EC plot and Fig. 4.24 shows the $\delta^{18}O$ – Tritium plot. Both confirm the absence of interconnection between the surface waters in the metasedimentary & igneous formation and the karstic aquifer. Most of the groundwaters had tritium in the range of 4 to 10 TU indicating modern recharge. The tritium variation was attributed to the anisotropy of the karstic aquifer. The spring had minimum tritium level at 0.5 TU possibly indicating a long and tortuous path in the metasedimentary & igneous rocks.

The dug well samples showing high nitrate content also showed high tritium content. High nitrate levels could be the result of irrigation return flow.

Conclusion

The study showed that

a) the groundwater in the karstic aquifer was mainly recharged directly by rainfall and the recharge appeared to be rapid

b) there was no interconnection between the reservoir located at a higher altitude in the metasedimentary & igneous rocks and the karstic aquifer

c) the spring water in the hill appeared to be the oldest water body indicating the groundwater flow in the metasedimentary rock was slow

4.2.3 Arid Zone Ground waters (Thar Desert)

Isotopes are particularly suited to recharge studies on arid zone ground waters because

- Deep ground waters in desert areas are usually very old and identifiable by carbon-14 levels
- Tritium is usually absent in these waters
- Palaeowaters have distinct isotope signatures
- Present day rainfall is generally enriched in heavier stable isotopes compared to earlier pluvial episodes since, as mentioned earlier, the rain drops in arid environment are prone to evaporation during their fall through relatively dry air.

Following are some examples of groundwater studies carried put in the Thar Desert in Rajasthan, India (Fig. 4.25)

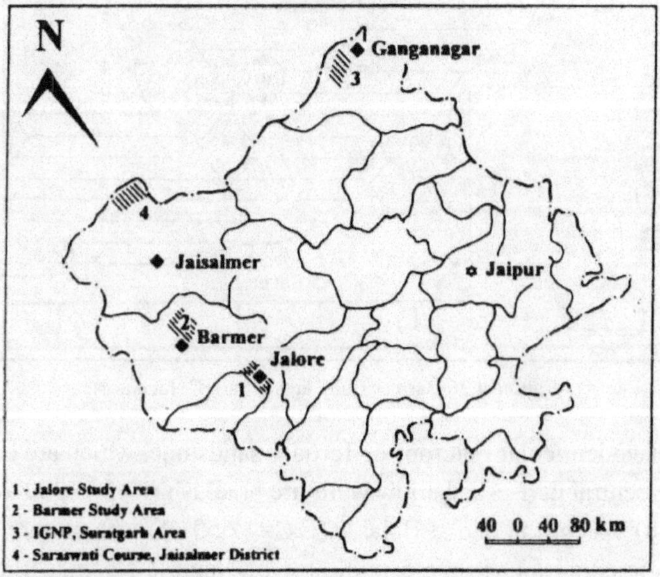

Figure 4.25. Map of Rajasthan indicating sites of isotope studies

4.2.3.1 Barmer Study [16]

Barmer district is in southwestern Rajasthan and has a mean annual rainfall of ~280 mm. Fig. 4.26 shows the study area (~100 km^2) and its geology.

The Tertiary aquifer has relatively fresh water (Nagurda, Bheemda, Nimla with E.C < 2000 μS/cm) and is of particular interest as it is surrounded by saline water formations. Lathi formation of Jurassic age is on the north and Malani suite of igneous rocks is present on the other sides.

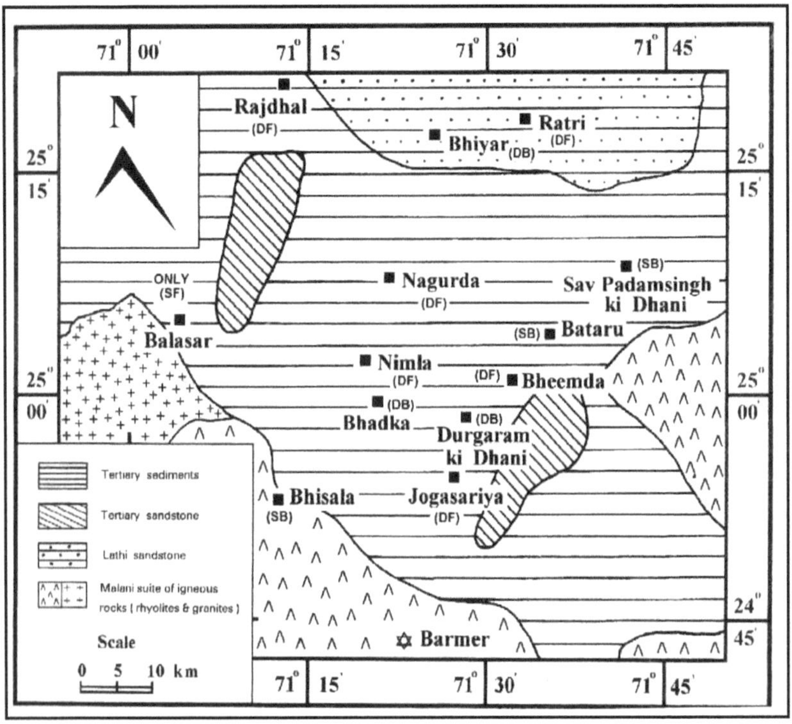

Figure 4.26 Barmer study area – Sample locations

Two lenticular outcrops of Tertiary sandstone, which are dry, are in the central part. All shallow wells are brackish (E.C – 3000 to 7000 μS/cm) except the one at Balasar (E.C= ~650 μS/cm) and are under phreatic condition. In the central area of Nagurla-Neemla-Bheemda fresh water forms a basin leaving a narrow valley on the eastern and western sides. Generally the deeper aquifers, both fresh and brackish, are under semi-confined or confined condition.

The isotope study was intended to understand the recharge process. The following table provides the isotope and other data

Sample Location	Depth (m)	E.C (μS/Cm)	δ^2H (‰)	$\delta^{18}O$ (‰)	^3H (TU)	^{14}C (pMC)	^{14}C age (αBP)
Deep Wells							
Bheemda	280	1830	-50.1	-7.30	0.0	22	7400
Jogasaria	285	2450	-51.7	-7.70	1.4	25	5500
Bhadka	200	4500	-51.5	-7.47	0.7	-	-
Nimla	220	1560	-56.1	-7.98	1.0	50	2400
D'ki'Dhani	100	3850	-43.2	-6.00	0.9	-	
Rajdhal	125	1380	-53.4	-8.10	1.7	-	
Bhiyar	100	3000	-36.3	-5.20	4.5	-	
Ratri	100	1820	-35.9	-5.19	-	-	
Nagurda	95	1710	-53.4	-8.00	-	-	
Shallow Wells							
Bheemda	-	6700	-40.8	-5.60	6.0	-	-
Bataru	-	5600	-34.1	-4.45	3.0	-	-
D'ki'Dhani	-	4450	-45.1	-6.48	3.0	-	-
Balasar	-	630	-20.3	-4.19	21.0	-	-
S'P'Ki'Dhani	70	3200	-38.9	-5.03	2.9	-	-
Bhisala	40	4400	-26.9	-3.95	-	-	-

Fig. 4.27 shows the δ^2H - $\delta^{18}O$ plot for the Barmer samples. Fresh deep groundwater samples are depleted in heavier isotopes and form Group A along MWL. These samples show low tritium levels and are in the central groundwater basin. Brackish samples, both shallow and deep, fall in a separate group, Group B. These samples appear to have undergone some evaporation. All shallow and some deep brackish samples in B show measurable tritium (up to 6 TU) indicating some component of recent recharge.

A shallow sample from Balsar (situated near igneous outcrops on the southwestern side), which is fresh, showed modern tritium level of 21 TU (value relevant to precipitation in the eighties). This well probably taps water from the weathered igneous rocks.

The deep fresh ground waters showed carbon-14 model ages of 2400 years for Nimla, 5500 years for Jogasariya and 7400 years for Bheemda. These waters appear to have been recharged in the cooler and pluvial phases in the Holocene. If the Malani igneous rocks are considered to be the recharge zone, a ground water velocity of 6 to 10 m/a appears probable.

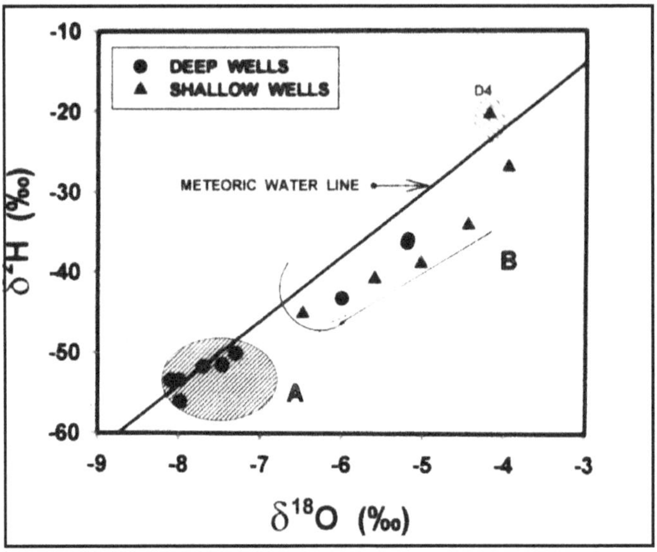

Figure 4.27 δ^2H - $\delta^{18}O$ plot of Barmer samples

4.2.3.2 Jalore Study [21]

Situated southeast of Barmer, Jalore district receives a mean annual rainfall of ~380 mm and is drained by the ephemeral Sukri River flowing southwest towards the Rann of Kachchh in Gujarat. Fig. 4.28 shows the study area and its geology. The area is extensively covered by alluvium. Sand dunes are a common feature especially in the western part.

The younger alluvium is mostly present along the Sukri River course and comprises unconsolidated to semi-consolidated coarse-to-fine sand and gravel.

The older alluvium of sub-recent to the Pleistocene age consists of semi-consolidated to consolidated coarse-to-medium sand with clay lenses, caliche and rock fragments. Tertiary sediments, consolidated clays with sand and shale fragments are encountered at depth in the western and southwestern parts.

Shallow groundwater mostly occurs under phreatic conditions. Confined to artesian flowing conditions exist in the southwestern part. An artesian flowing well is located at Jodhawas, which mainly taps the Tertiary aquifer.

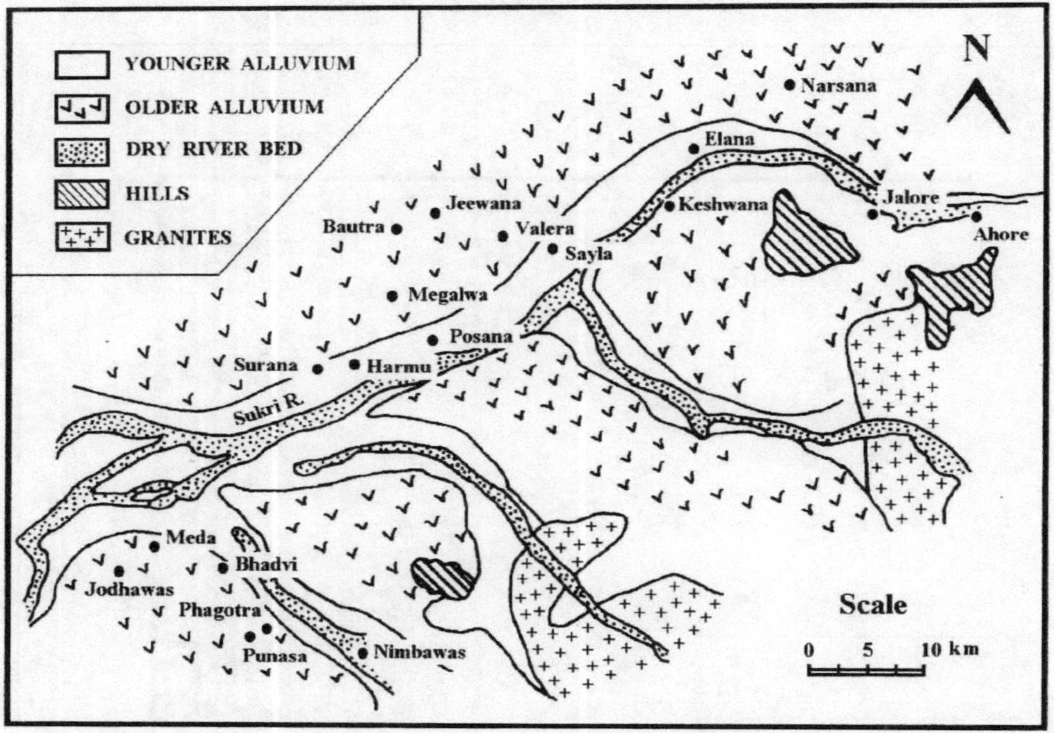

Figure 4.28. Jalore study area and sample locations (·)

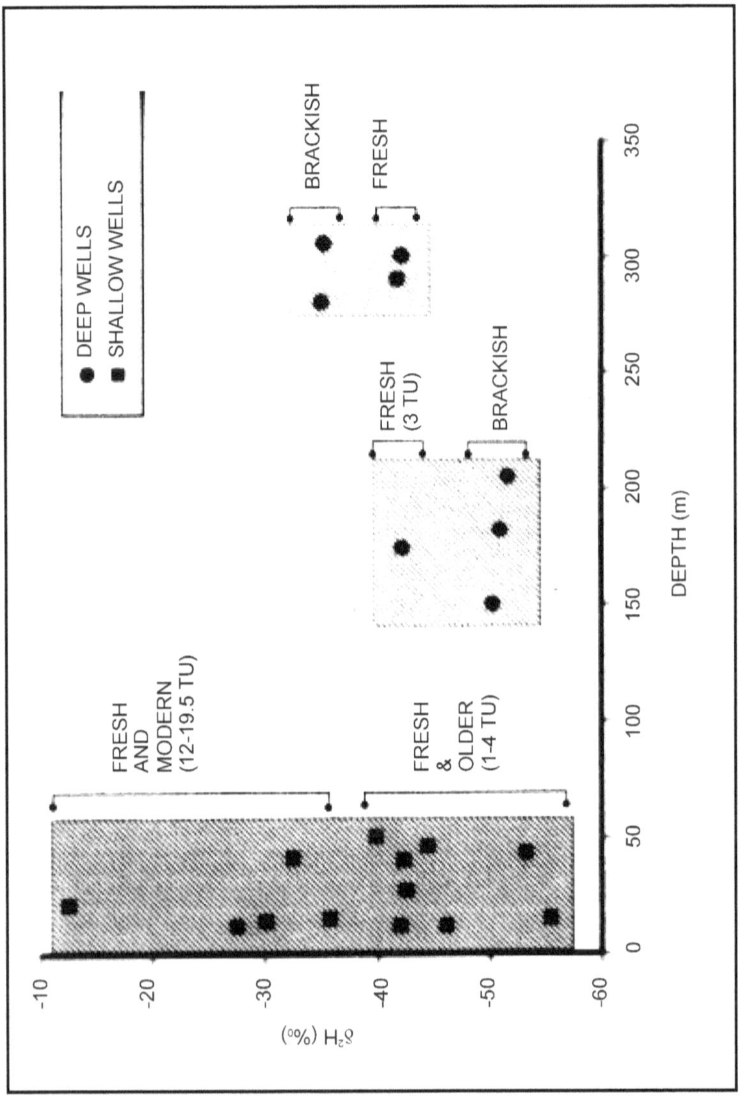

Figure 4.29 δ^2H vs. Depth (Jalore samples)

A number of samples from shallow (< 50 m) and deep (> 150 m) zones were analyzed for stable isotopes, tritium, cabon-14 and chloride and presented in the following table.

In Fig. 4.29 the three aquifers show different groups of isotope signatures. The deep aquifer at 300 m shows enriched stable isotope content compared to the brackish waters of the intermediate aquifer (150-200 m).

These are probably the waters of the last pluvial episode leaching soil salts left behind by earlier aridity. It is interesting to note that the fresh waters in all the three aquifers have similar isotopic composition. Shallow aquifer has waters of variable isotope composition with measurable tritium.

Place	Aquifer	Chloride (mg/L)	Tritium (TU) ó = +/- 0.5TU	C-14 (pMC)	δ^2H (‰)	δ^{18}O (‰)
Bautra	Deep	170	0.8	-	-41.8	-5.8
	Shallow	160	1.9		-39.8	-6.5
Megalwa	Intermediate	800	0.6	-	-51.6	-6.8
	Shallow	950	3.0		-42.0	-6.6
Punasa	Deep	260	2.4	-	-42.2	-6.1
	Shalow	160	4.0		-44.4	-5.8
Jodhawas	Deep	1360	1.2	12	-35.3	5.1
Meda	Deep	1330	1.6	-	-35.1	-5.5
Posana	Intermediate	780	0.9	-	-50.9	
Phagotra	Intermediate	-	-	-	-50.2	-6.7
Sayala	Intermediate	230	3.0	71	-42.2	-6.4
	Shallow	-	-	-	-55.4	-7.3

Fig. 4.30 shows the δ^2H - δ^{18}O plot of Jalore samples. Samples, both shallow and deep, appear to fall isotopically into three groups.

Shallow waters: Groundwaters along the river course show evaporative enrichment of heavier isotopes and fall in group C. They show modern tritium levels of >5TU. Groundwaters away from the river course show comparatively depleted heavier isotope composition and fall in group B. They have tritium values <5TU. Group A represents shallow waters with most depleted heavier stable isotope values and no detectable tritium.

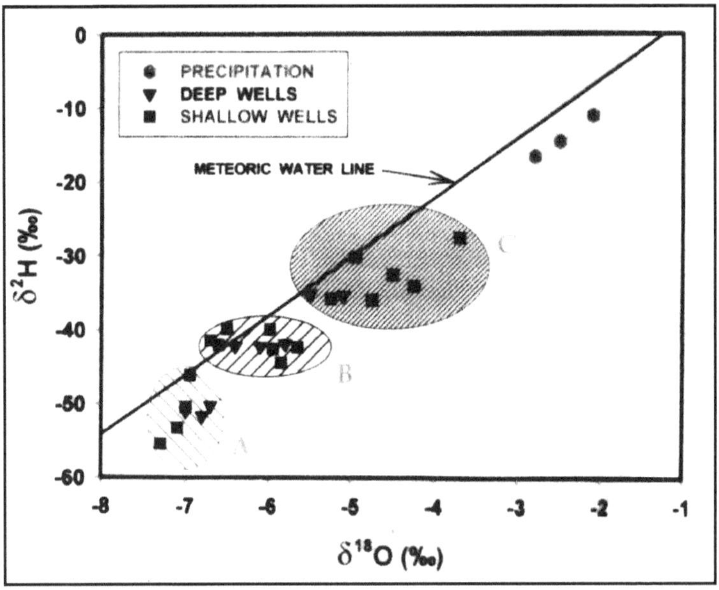

Figure 4.30 δ^2H - $\delta^{18}O$ plot of Jalore samples

***Deep waters*:** Deep well samples also seem to fall into three groups as above. The most depleted samples (Group A) are from intermediate depths (Megalwa, Posana & Phagotra), are brackish and have no detectable tritium indicating absence of any recent recharge. Group B (Baura, Punasa & Sayla) ground waters, which are close to the river course, are slightly enriched in heavier isotopes, are fresh and show measurable tritium. They appear to receive recharge from shallow groundwater. Deep well at Sayala appears to tap both shallow and intermediate aquifers as is evident from stable isotopes, tritium and carbon-14 values. The shallow and deep aquifers at Bautra and Punasa appear to be interconnected as they have similar stable isotope composition (see Fig. 4.29) and similar piezometric levels (not shown in the table). The deep well at Jodawas is artesian flowing type, brackish, has no detectable tritium and has a carbon-14 level of 12 pMC, corresponding to an uncorrected age of 17,500 years.

The study indicated that the shallow aquifer receives recharge through river channels during episodic floods. In some areas, it may also be receiving recharge from the intermediate confined aquifer by upwelling through subsurface fault zones. The deep aquifer, however, does not appear to be receiving any modern recharge. It was probably last recharged during pluvial phase in the early Holocene. The intermediate aquifer (150-200 m) was probably recharged during the cooler, humid period in the mid Holocene.

4.2.3. Study on a Palaeo buried Channel of Sarasvati(?) in Jaisalmer district

Background

There is a large interest in studies on palaeochannels in the Thar Desert, which were river channels buried several thousands of years back with the onset of aridity in Western Rajasthan. Remote sensing and field observations have recently identified several such palaeochannels [22]. One of them belongs to the legendary Sarasvati River.

Ample evidence exists from ancient Indian texts, geomorphology and sedimentology (see Bibliography No.3) that a mighty river once originated in the Himalayas flowed in the Northwest India during 7000-3000 BP and disappeared in the Rajasthan desert. Rigveda describes the birth and glory of the Sarasvati River. In their praise of God Indra for killing the serpent demon 'Ahi' and releasing the waters held by the demon they say " Oh Indra! On your being born and with fear of your rage heavens trembled, huge mountains were fearful and river waters started flowing moistening the desert. Oh Indra! You struck down barriers and broke open mountains". Killing of serpent demon 'Ahi' may mean melting of a glacier and reference to breaking up of mountains may refer to tectonics. Which one of these is responsible for the birth of Sarasvati? Rigvedic description of the sudden appearance of the river and 'its majesty and tempestuous roar while cascading down the Himalayan slopes and reaching the ocean' distinctly points out tectonics being responsible for the birth of Sarasvati. Isotopes help to confirm this, as we shall see later.

In fact, Rigveda mentions release of seven rivers; Sarasvati, Sutlej, Chenab, Ravi, Beas, Jhelum and Indus (all present day names). Whereas all the others are still major rivers in Northwest India and Pakistan, Sarasvati dwindled down to a small monsoon rivulet and is a mere tributary to a small river Ghaggar that disappears into the sands of the Thar Desert. The epic Mahabharata refers to it as a sluggish river,

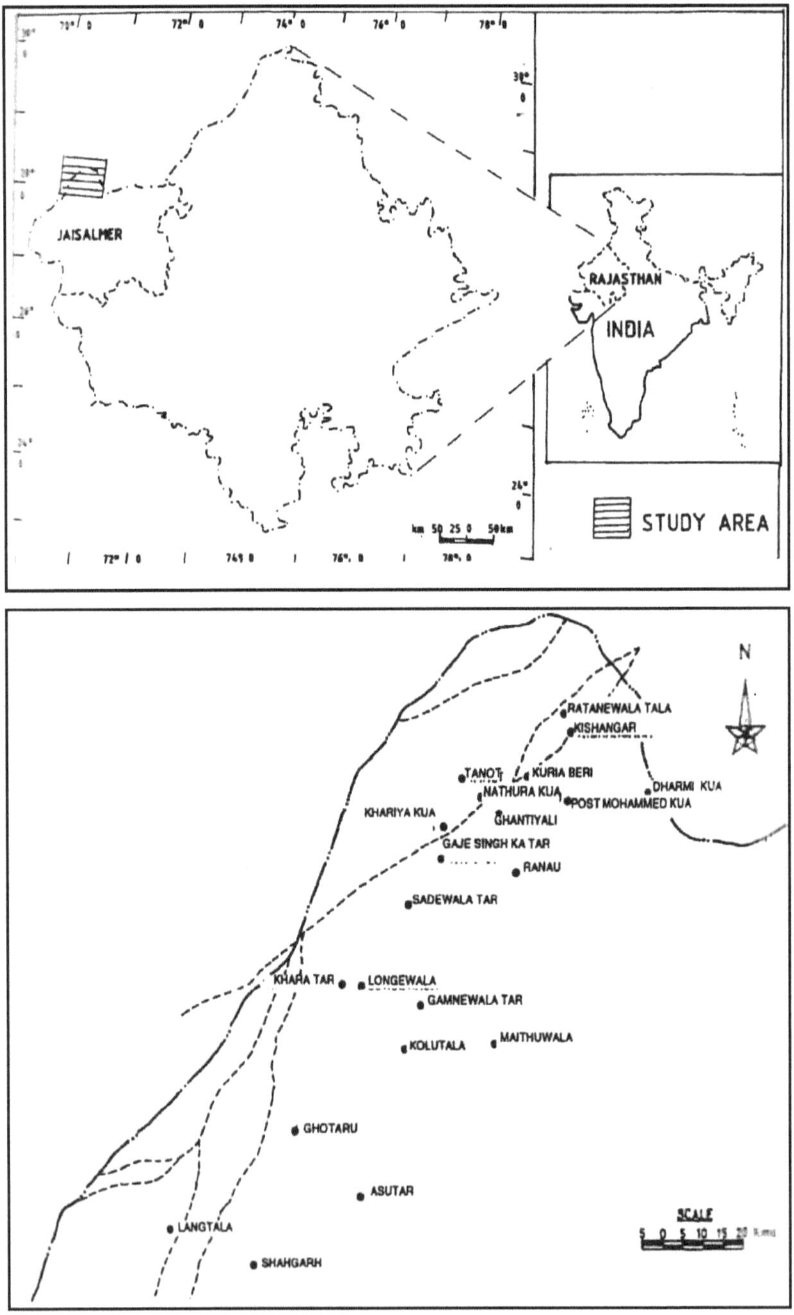

Figure 4.31 Location map of the Sarasvati study area

meandering and migrating. The Puranas give its place of birth as Plaksa prasravana (lower Himalayas) and disappearance as Vinasana (Rajasthan). There are similar to the source and place of disappearance of the present day Ghaggar River. The Puranas also refer to its flow through a series of lakes towards the end of its life.

Though the wet climate ended around 4000 BP in large parts of Northwest India, it does not fully explain the disappearance of Sarasvati. A combination of tectonics around 3000 BP cutting off the main tributaries, Yamuna and Sutlej and onset of aridity might have caused the disappearance. Search has been going on for over a century to locate the buried channel of the Lost River.

The Ghaggar-Hakra-Nara dry riverbed extending from Ganganagar district in Rajasthan, entering Pakistan and reentering India at Tanot in Jaisalmer district attracted special attention. Remote sensing [23] detected two courses, one westward to Fort Abbas in Pakistan and the other towards Khangarh in the southwest, also in Pakistan. It is the latter that reenters India at Tanot. Here again it has two branches, one north of Tanot and the other 14 km south of Tanot. It is the latter, which generates maximum interest. The specific course of interest beyond Tanot is along Kishangarh, Ghantiyali and Shahgarh (Fig. 4.31)

In spite of the highly arid conditions, comparatively good quality of groundwater is available along the above channel at 30 m depth. A few dug wells along the course do not dry up in summer and the tube wells do not show any reduction in the water table even after extensive exploitation. All the above suggested that the Kishangarh-Ghantiyali-Ghotaru course could be part of the buried channel of the ancient Sarasvati River. This region was thus selected for isotope study.

Isotope study

The objective of the isotope investigation [20,24] of the above mentioned course was to characterize the shallow and deep ground waters, determine their ages, identify the sources of recharge and confirm headwater connection if any in the Himalayas.

Several samples along the selected course of the buried channel have been analyzed for stable isotopes, carbon-14 and tritium as well as for electrical conductivity. The following table gives the data for some selected samples.

Location	*DW/TW/HP	EC (μS/cm)	δ ¹⁸O ‰	³H (TU)	¹⁴C age (αBP)
Kishangarh	TW	3500	-5.6	n.d	6000
Kuriaberi	DW	2100	-5.7	n.d	4500
Nathura kua	DW	3000	-6.3	n.d	3000
Ghantiyali	TW	3700	-6.6	n.d	9500
Ghantiyali	DW	2800	-6.0	n.d	5000
Ranau	TW	1900	-6.2	n.d	6000
Longewala	TW	2740	-6.2	n.d	19000
Ghotaru	TW	2300	-6.9	n.d	13000
Ghotaru	DW	3700	-6.4	1	4000
Asutar	TW	2600	-6.3	n.d	13000
Asutar	DW	2400	-	n.d	-
Langtala	HP	3400	-5.0	n.d	3000
Langtala	DW	2380	-6.0	1	3500
Dostmoh kua	DW	1400	-6.5	1	6000

*DW/TW/HP refer to Dug well/Tube well/ Hand pump

n.d= non-detectable

Water table contours indicate groundwater flow in the NE-SW direction. Hydrogeologically both shallow and deep waters are in the same regime, the tube wells representing the deeper horizon. The tube wells have screens at varying depths in the range of 73 to 150 m below ground surface. The shallow and deep waters have similar EC and stable isotope composition. In the δD - $\delta^{18}O$ plot in Fig. 4.32, it is seen that the $\delta^{18}O$ values cluster around –6‰ and the tube well samples fall on an evaporation line, which intersects the Meteoric Water Line (MWL) at the $\delta^{18}O$ value of about –9‰. This source value of $\delta^{18}O$ is slightly enriched compared to other Himalayan rivers of glacial origin like Chenab (-12 ‰, see 3.6.3) and Ganga (-10 ‰, see 3.2.1).

The tube wells and most of the dug wells have no detectable tritium indicating absence of modern recharge. Some dug wells do show traces of tritium showing some component of recent recharge. Carbon-14 values indicate uncorrected ages of thousands of years; dug wells: 3000 to 6000 years and tube wells: 6000 to 19000 years. There is a trend of increase in the apparent age of groundwater from Kishangarh to Ghotaru indicating a velocity of less than 5 m/year.

From the above, it may be said that the isotope data when interpreted with remote sensing and ground data as well as descriptions in ancient Indian texts appears to support the view that the palaeochannel in the Kishangarh – Ghantiyali – Ghotaru sector could be part of the lost Sarasvati River. The ground waters are thousands of years old and appear to move slowly in the aquifer system. The aquifer was probably recharged when the ancient river was flowing. No modern recharge is evident. But, the fact that the groundwater levels remain steady even after copious exploitation shows continued headwater connection. The stable isotope composition indicates that the lost river had originated at an altitude, probably in the lower Himalayas.

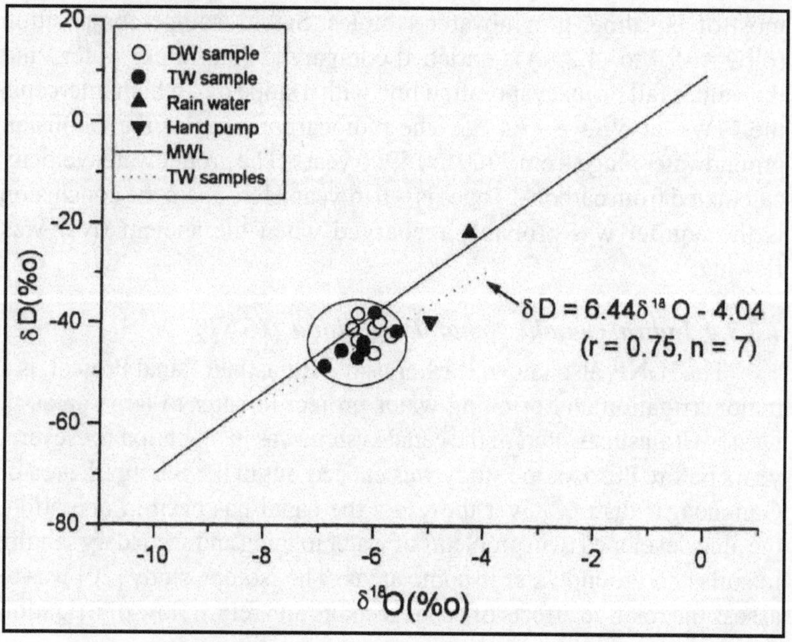

Figure 4.32 δD - δ^{18}O plot of Sarasvati samples

The source δ^{18}O value of –9‰ clearly indicates that the Sarasvati River could not have been of glacier origin, but the result of the alternate possibility of tectonics. The Rigvedic description of the birth of Sarasvati can now be more correctly interpreted. The Puranic description that the river flowed through a series of lakes towards the end of its life is reflected in the δ^{2}H - δ^{18}O plot in Fig. 4.32 with the tube well samples falling on an evaporation line with a slope of 6.4.

Isotope study on the branch entering Pakistan

As mentioned earlier, the dry Ghaggar – Hakra – Nara riverbed enters Pakistan, divides into two courses; one course is traced westward towards Fort Abbas and the other towards Khangarh and reenters India. The westward course was similarly investigated using isotope techniques by German researchers [25] in the Fort Abbas – Mojgarh sector in the Cholistan part of the Thar Desert. This part of the desert receives an annual rainfall of about 200 mm. Here also fresh groundwater occurs at a depth of 50 m with an aquifer thickness of 100 m sandwiched by brackish to saline water. Groundwater has no detectable tritium like most of Jaisalmer groundwater samples. Stable isotope composition ($\delta^{18}O$ = -5.7 to -4.2‰) is enriched compared to Jaisalmer waters and the values fall on an evaporation line with a slope of 4 which intercepts the MWL at $\delta^{18}O$ = –7.4 ‰. The radiocarbon ages of the Cholistan groundwater range from 7700 to 15900 years. The groundwater velocity calculated from carbon-14 ages is ~ 6 m/year. Here again the conclusion is the aquifer was probably recharged when the ancient river was flowing.

4.2.3.4 Indira Gandhi Nehar Pariyojana (IGNP)

The IGNP, also known earlier as the Rajasthan Canal Project, is a major irrigation and drinking water project to cater to large areas in western Rajasthan. Part of the canal system was in operation for several years before this isotope study was carried out in the Suratgarh area of Ganganagar district. Over the years the canal has been in operation, the area developed twin problems of water logging and secondary salinity in soils and groundwater in some areas. The isotope study [20] was to assess the relative effects of canal seepage and return flow of irrigation to the observed problems.

Fig. 4.33 gives the δ^2H - $\delta^{18}O$ plot of the samples collected from the canal and from groundwater. The canal waters originating from the Bhakra-Nangal dam in Punjab are Himalayan waters and hence depleted in stable isotope composition compared to local precipitation and groundwater. The canal waters fall in Group A. Groundwater samples in Group B fall on a line nearly parallel to MWL and influenced only by seepages from the canal. Peizometer samples affected by return flow of irrigation, water logging and saline groundwater fall in Group C with evaporative isotope signature of slope 5.

This was to be an exploratory investigation to be followed by a more detailed study. No such study could be made. In any case, this example is demonstrative enough of the potential of isotope techniques.

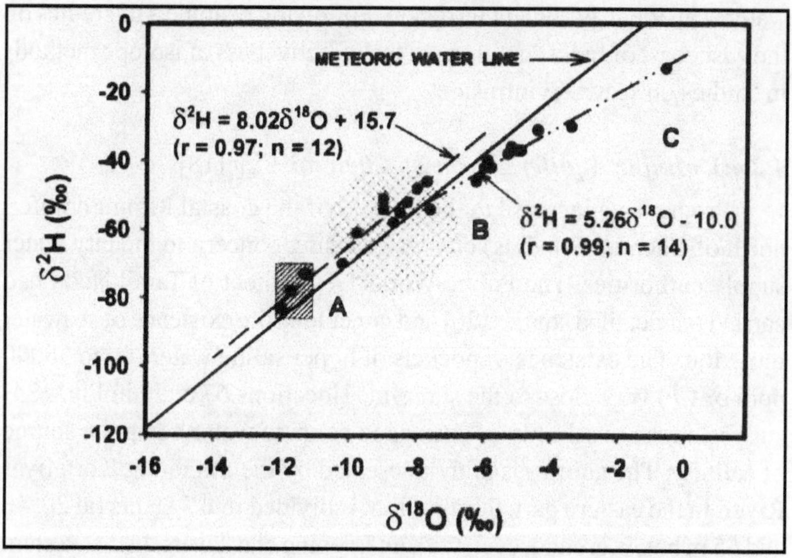

Figure 4.33 δ^2H - $\delta^{18}O$ plot of canal (IGNP), groundwater samples at Suratgarh

4.2.4 Coastal Aquifers – Seawater Intrusion

Coastal aquifers are one of the end points of the hydrological cycle. In natural conditions hydrostatic balance exists with the lighter fresh water floating over the denser seawater. As human settlements along the coast are usually large and so is the exploitation of coastal aquifers, the natural fresh water-seawater balance is often disturbed with the possible seawater intrusion.

Degree of influence of seawater on a coastal aquifer depends on

- ❑ Local hydrology and hydrogeology
- ❑ Level of exploitation of the aquifer
- ❑ Climate changes in the geological past

Broadly two types of seawater intrusion is usually encountered

1. Present day active intrusion
2. Palaeo-intrusion entrapped in foreshore sediments and contaminating the overlying fresh water aquifers

Isotope hydrology can play a distinctive role in studies on seawater intrusion as the stable isotope composition and chemistry of fresh water and seawater are different. Levels of tritium and carbon-14 in coastal waters can identify ancient intrusion. Following examples of studies on the east coast of India demonstrate the effectiveness of isotope methods in studies on seawater intrusion.

4.2.4.1 Minjur Aquifer, North of Chennai City [18]

Progressive increase in the salinity of the coastal Minjur aquifer, north of Chennai (Madras) city was causing concern to the city water supply authorities. The Public Works Department of Tamil Nadu had carried out detailed studies [26] and concluded the existence of seawater intrusion. The existence of pockets of hyper saline water (up to 36000 ppm of Cl^-) very close to the shoreline (locations 67 & 70 in Fig.4.34) suggested the possibility of entrapped connate water being the source of salinity. The aquifer is alluvial created by the migrating Kortaliyar River. In the eastern part, the alluvium is divided into 3 layers (at 20, 40 and 55 m below ground level) with intervening clay layers. In the western part where the well field is situated, the three zones merge into one.

Saltpan activity (production of salt from seawater) was popular on the coast for many decades till the seventies.

Isotope and chemistry data

On the δD - $\delta^{18}O$ plot (see fig. 4.35), the saline samples fall on a mixing line between seawater and the production wells (fresh water).

Following is a comparison of the chemistry and tritium levels of seawater and the hyper saline waters

Sample No.	Chloride (ppm)	I^-/Cl^-	Br^-/Cl^-	Tritium (TU)
67 (HS)	28700	2.6×10^{-6}	1.8×10^{-3}	2.8
70 (HS)	36000	1.25×10^{-6}	1.25×10^{-3}	-
Seawater	18000	2.9×10^{-6}	2.8×10^{-3}	6.7

HS: Hyper-saline

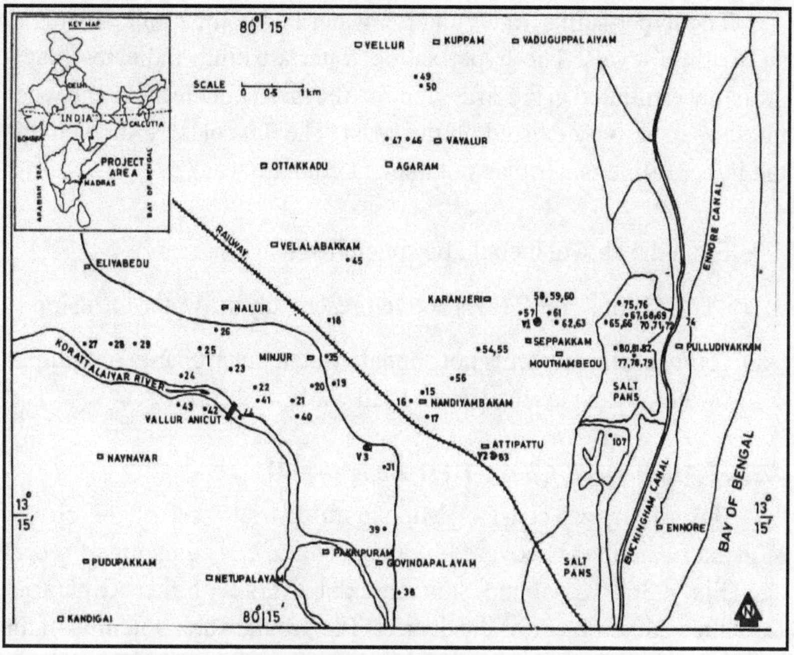

Figure 4.34 Minjur aquifer study area with sample locations

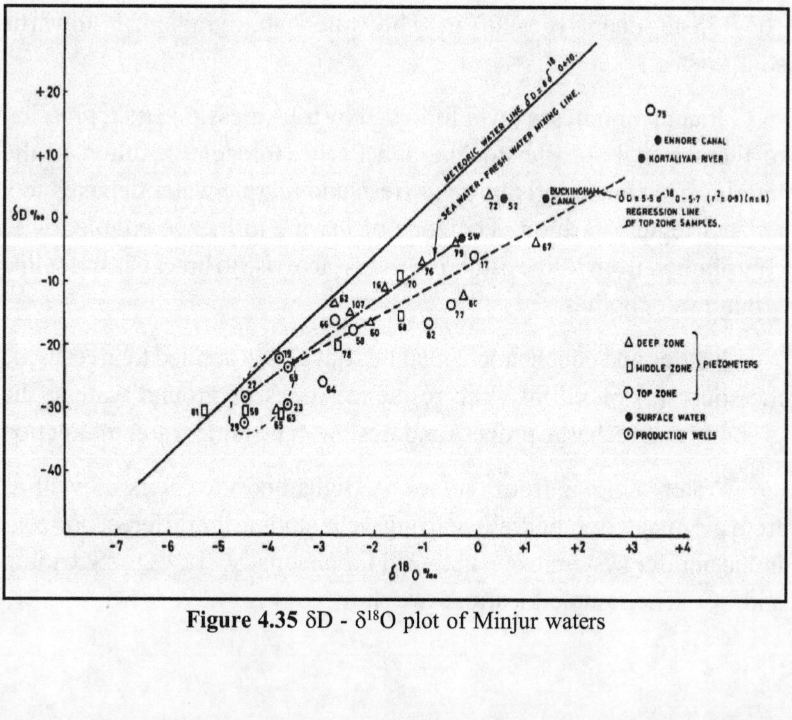

Figure 4.35 δD - δ^{18}O plot of Minjur waters

The hyper-saline water and Seawater had comparable chemistry and tritium levels. The hyper saline water's tritium indicates that it probably infiltrated in the fifties, before the thermonuclear era and when the saltpan activity existed on the coast. The data clearly showed that the hyper saline water was not ancient connate water.

From the above, it could be concluded that

- The Minjur aquifer has active present day seawater intrusion
- Hyper saline water is not connate water, but probably old saltpan water, which leached excess salt while infiltrating.

4.2.4.2 Midnapore Coast, West Bengal [27]

Bay of Bengal coast of Midnapore district west of the city of Kolkata (Calcutta) is covered by unconsolidated sediments of recent age (Fig. 4.36). Kosai and Subarnarekha rivers with their tributaries form the main drainage of the district. The groundwater potential of the unconfined/semi-confined aquifer, though good, is beset with problems of salinity. The aquifer is a 3-tier system; shallow: 35-100 m, middle: 105-155 m, deep: 165-200 m. The water table gradient is towards southeast.

Fluctuation in sea level followed by transgression and regression of the sea in Late Pleistocene and Early Holocene resulted in the interlayering and overlapping of fresh and marine water deposits in a rather irregular fashion. The zone of marine influence coupled with contribution from saline estuarine rivers gave rise to brackish and saline groundwater bodies.

Isotope and geochemical methods have been applied to investigate the source(s) of salinity, the residence times of ground waters, the groundwater recharge process and fresh water- saline water interaction

Water samples from the sea at Digha and the rivers as well as from piezometers, tube wells and dug wells tapping the different horizons in the aquifer system were analyzed for chemistry, ^{2}H, ^{18}O, ^{34}S, tritium and ^{14}C. The sample locations are shown in Fig. 4.37.

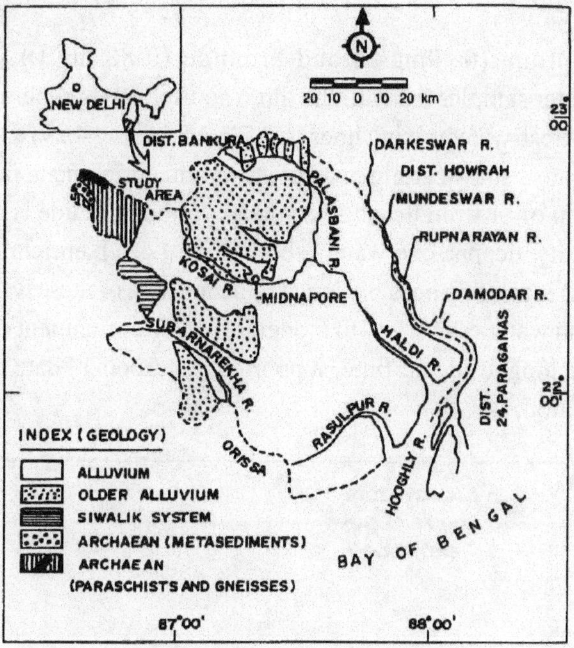

Figure 4.36. Geological map of Midnapore

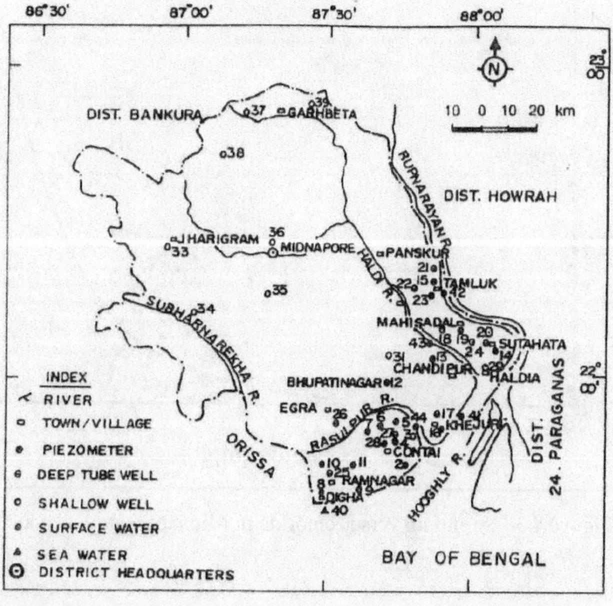

Figure 4.37. Location of sampling points, coastal Midnapore

Chemistry

Strontium (0- 9mg/L) and bromide (0-80 mg/L) values of groundwater samples from all the three aquifer zones generally fall on seawater-fresh water mixing lines (see Fig.4.38 & Fig.4.39). Deep saline ground waters show certain level of strontium enrichment, possibly by dissolution of Sr from the aquifer matrix. Though iodide is correlated with salinity, deep saline waters show (Fig. 4.40) I^- enrichment (250-550 mg/L) indicating possibly long residence time of these waters. Even some fresh waters show low to moderate iodide enrichment once again indicating long residence time, supported by carbon-14 data given later in this section.

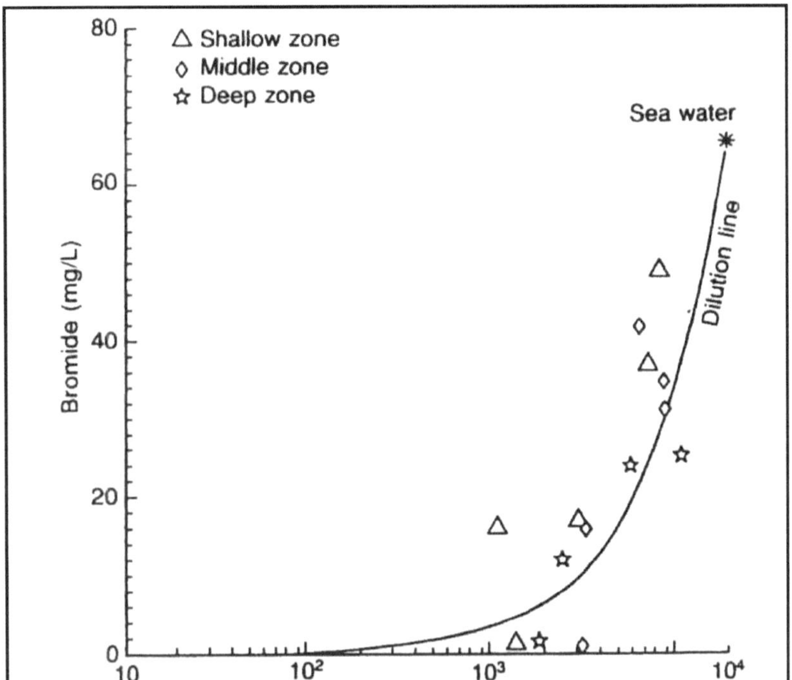

Figure 4.38 Strontium versus chloride in Midnapore ground waters

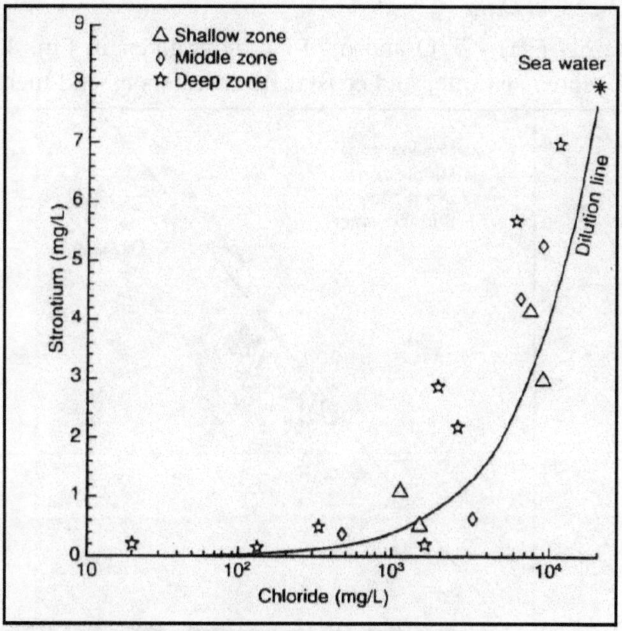

Figure 4.39. Bromide versus chloride in Midnapore ground waters

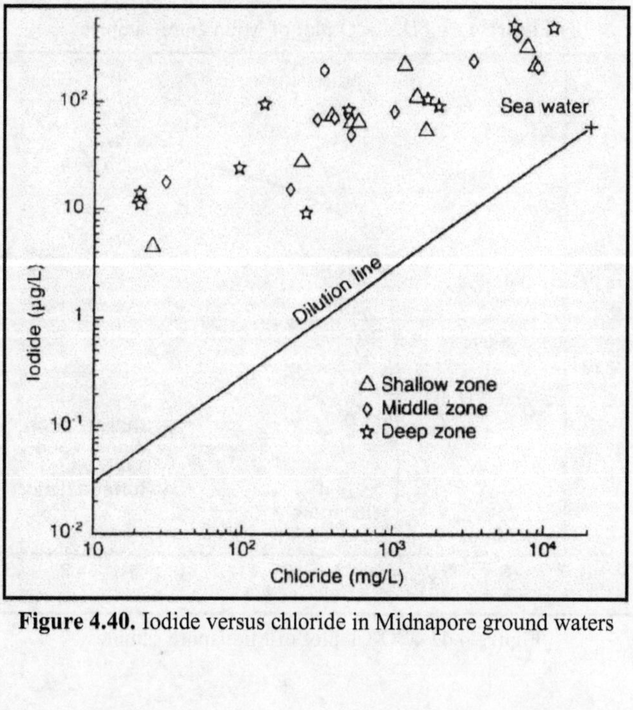

Figure 4.40. Iodide versus chloride in Midnapore ground waters

Stable Isotope Data

The δD (^2H) - $\delta^{18}O$ and $\delta^{18}O$-Cl^- plots given in Fig. 4.41 and Fig. 4.42 show seawater and coastal fresh waters as end members.

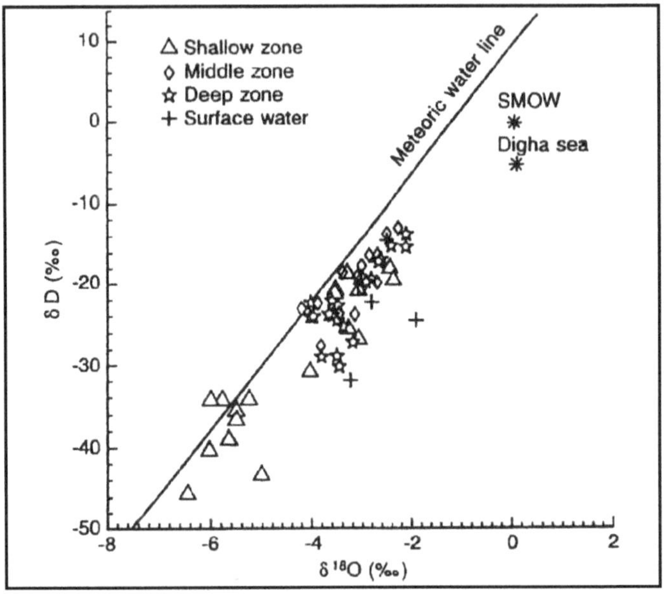

Figure 4.41 δD - $\delta^{18}O$ plot of Midnapore samples

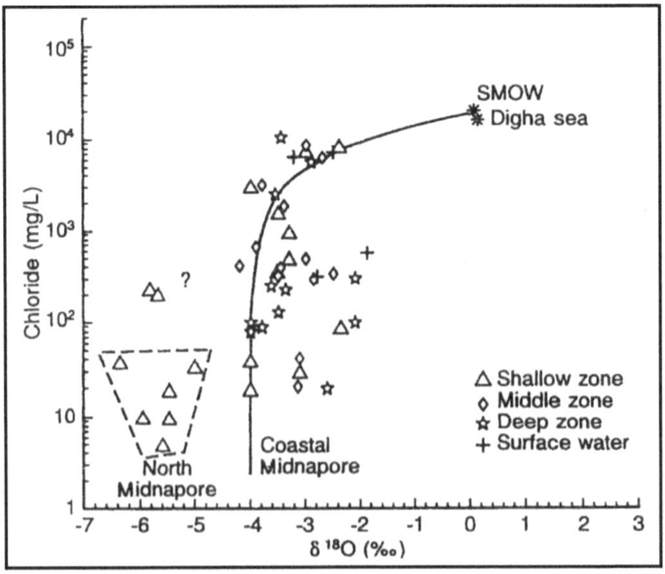

Figure 4.42 $\delta^{18}O$-Cl^- plot of Midnapore samples

Saline surface waters of the coastal rivers show evaporative enrichment and fall away from the mixing line. North Midnapore fresh shallow waters fall as separate group in both the plots. Many of the middle zone samples of coastal Midnapore and quite a few of the deep zone saline and fresh waters fall on the seawater mixing line.

Following table shows the $\delta^{34}S$ values of some selected groundwater samples and of seawater

Source	Depth (m)	Sulphate (mg/L)	$\delta^{34}S$ (‰, CDT)
Sarada Seed Farm (4)	52	100	+31
Sarada Seed Farm	92	102	+30
Sarada Seed Farm	176	110	+32
Marisda (5)	161	57	+29
Marisda	192	192	+29
Digha seawater			+21

* Numbers in brackets refer to sample locations given in Fig. 4.37

Whereas the seawater exhibited a $\delta^{34}S$ value of +21‰ (close to a generally accepted value of ~ +20‰ for seawater), the shallow and deep water samples from Sarada seed farm and Marisda (Southeast of Midnapore towards the coast), showed enriched values of +29 to +32‰ and at these places H_2S could be smelt. Obviously, the enrichment is due to bacterial reduction of sulphate to sulphide under reducing conditions, which, once again indicated long residence time.

Tritium and carbon-14

Tritium and carbon-14 data of some selected locations are given in the following table

Location	Depth (m)	Chloride (mg/L)	$\delta^{13}C$ (‰,)	^{14}C (pMC)	^{14}C age (a BP)	^{3}H (TU)
Sarada Seed Farm (4)	92	6300	-16.7	50	5600	1
Marisda (5)	162	8600	-15.3	53	4400	5
Nilpur (6)	105	500	-18.3	22	6700	2
Chandipur (13)	98	1500	-8.5	29	4600	5
Sutahata (14)	111	3300	-15.8	32	8700	1

* Numbers in brackets refer to sample locations given in Fig. 4.37

There is clearly a complex intermixing of very old saline waters with moderately old groundwater and recently infiltrated river waters. Generally saline waters are very old and do not represent modern seawater intrusion.

Conclusion

- The main source of salinity in coastal Midnapore groundwater appears to be seawater entrapped during Flandrian transgression in the Early Holocene (~8000 B.P).
- Shallow and middle zones are also affected by infiltration from saline rivers.
- The fresh water recharge, particularly to the deep zone, is from northwestern Midnapore outcrops.

4.2.4.3 Delang-Puri Sector, Mahanadi Delta, Coastal Orissa [28]

The Delang-Puri sector in coastal Orissa, India forms the Southwestern part of the Mahanadi delta (Fig. 4.43). Thousands of hand pumps installed in coastal Orissa, as part of a drinking water project in the 80's, quickly became unacceptable mainly because of progressive incursion of saline waters. This isotope geochemical investigation was part of a major hydrogeological study [29] on the recharge and salinization processes in the Delang – Puri sector.

The Mahanadi delta has a great aerial spread of 9000 km^2. On the landward side, to the west and northwest of Delang, the area is undulating upland with isolated hills and lateritic peneplains. The hills consist of the Eastern Ghats group (mostly khondalite, charnockite, leptynolite, granite gneisses etc.) of Archaean age.

A schematic representation of the subsurface geology along the Delang-Puri sector is shown in Fig. 4.44. It is seen that the basement, comprising the Archaean rocks deepens southward. The depth varies from 20 m at Jagadalpur to more than 600 m at Puri. The unfossiliferous yellow-brown sediment horizon (IIIB) is sandwiched between two dark grey fossiliferous sediment horizons (IV and IIIA & IIB) indicating a prominent change in environmental conditions during the evolution of the delta.

The Fig. 4.45 shows a complex multi-aquifer system in which fresh, brackish and saline waters occur in a rather complicated fashion. The shallow aquifers (depth <50 m) occur under phreatic conditions and deeper aquifers change from semi-confined to confined flowing type with increasing depth. A point of interest is the fresh water aquifer at a depth of 102-165 m with thickness ranging from 25 to 80 m. This fresh water aquifer is sandwiched between saline aquifers. Groundwater contours indicate Delang area to be the recharge zone and the general groundwater flow direction is southward to the sea.

The groundwater quality, in general, varies from fresh to very saline ranging from 40 μS/cm to 30,000 μS/cm. In the Delang area, shallow waters are generally fresh (EC ~1200 μS/cm and chloride ~200 mg/L). In general, the hydrochemical facies in the study area (Delang-Puri sector) change from $HCO_3 \rightarrow HCO_3 + Cl \rightarrow Cl + HCO_3 \rightarrow Cl$. The presence of H_2S has been detected at deeper levels at a number of places.

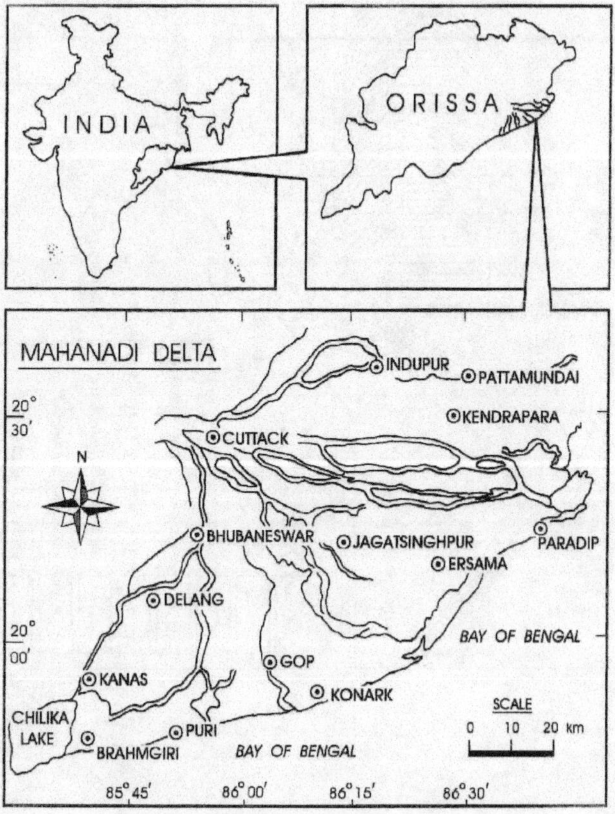

Figure 4.43 Mahanadi Delta, Study area

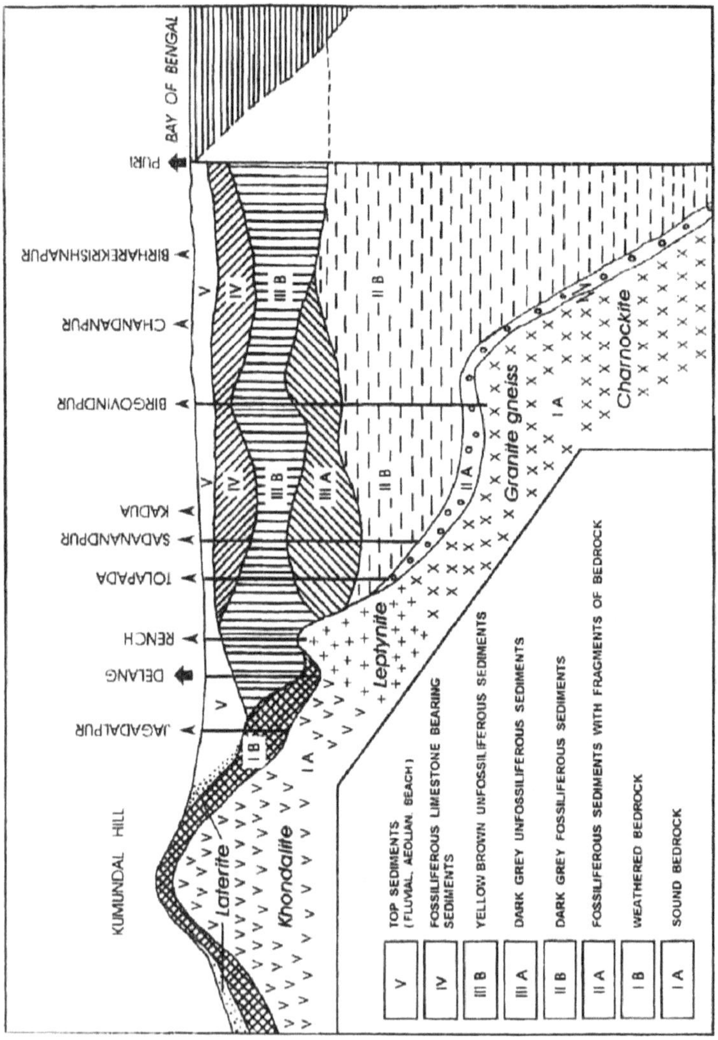

Figure 4.44. Subsurface geology along the Delang – Puri sector

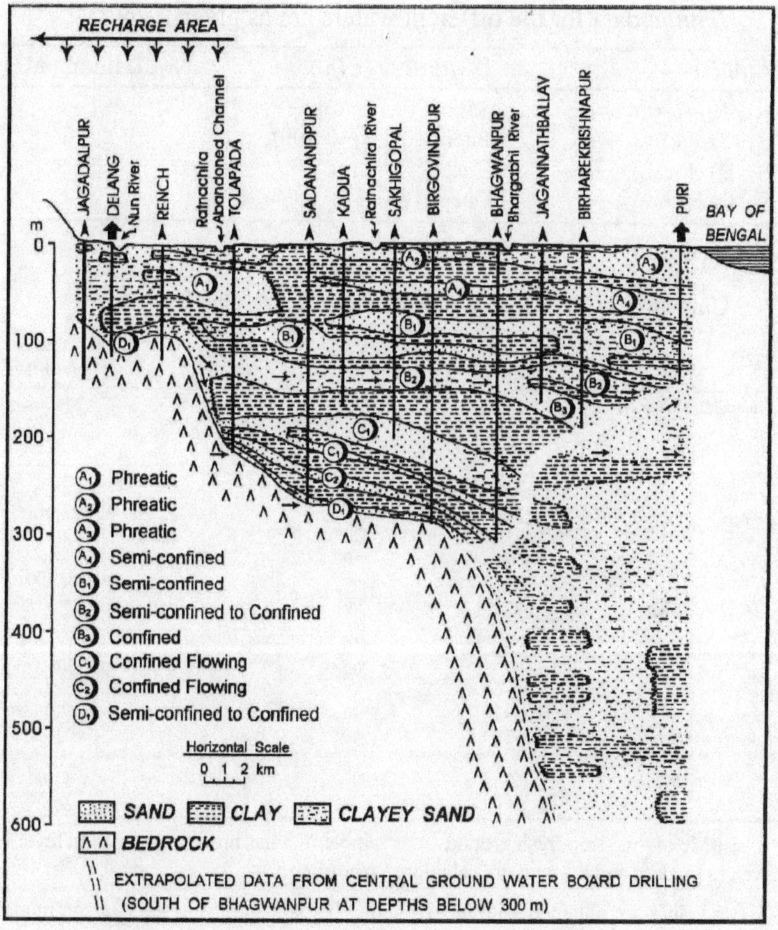

Figure 4.45 Multi-aquifer system along the Delang-Puri sector

Isotope Data

Fig. 4.46 shows the locations of sample collection for isotope analyses.

δD - $\delta^{18}O$ plots for several groups of samples are presented in Fig. 4.47. This figure is not available in Ref [28]

All fresh waters; surface, shallow and deep, show evaporative enrichment. Saline waters in the intermediate aquifers do not show evaporation whereas deep saline waters did undergo evaporation.

Tritium data for the different waters are as given below.

Aquifers	Depth Range (m)	Tritium (TU)
A_1, A_2, A_3 (Fresh)	Sallow (up to 50)	1 - 4
B_1 (Saline)	Intermediate (60-100)	n.d
B_2, B_3 (Fresh)	Deep (100 – 160)	3 - 8
C_1, C_2 Saline	Deep (180-250)	n.d

n.d = not detectable

Carbon-14 data of groundwater samples are as follows.

Place	Depth (m)	EC (μS/cm)	$\delta^{13}C$ (‰)	^{14}C age (a BP)
Gangapurpatna	16	780	-14.1	Modern
Sauria	40	540	-13.6	Modern
Rench	31	2230		4950
Rench	126	660		Modern*
Tolapada	21-25	530		650
Tolapada	47-53	560		880
Tolapada	187-190	3580		33,300
Prat'ram'pur	60	16200	-12.4	9850/7350#
Sadanandpur	125-128	2430		16450
Sadanandpur	Flowing	14000	-12.6	26550/24850#
Birgovindpur	184-193	5570	9.4	24450/19650#
Birgovindpur	240-293	21100	-17.0	46600/46350#
Birgovindpur	>250	29700		23100
Puri	115-118	12200		12550

*67.1 pMc of this deep fresh groundwater sample that has present day tritium level is taken as modern and ^{14}C ages are calculated accordingly

#These values are obtained with the available ^{13}C data and using the Pearson model (Eq.2.24)

Discussion

From the above data, we can draw the following information.

The shallow aquifers (A_1, A_2, A_3, A_4 in Fig. 4.45) with tritium levels of 1 to 4 TU are replenished by evaporated surface waters through laterites and outcrops of basement rocks exposed on hills northwest of Delang. In the southern plain the recharge could be infiltration of evaporated surface waters through flood plains, ponds and rivers.

The samples from the intermediate saline aquifer (B_1 in Fig 4.45) do not show any evaporation and are tritium-free indicating absence of any recent recharge.

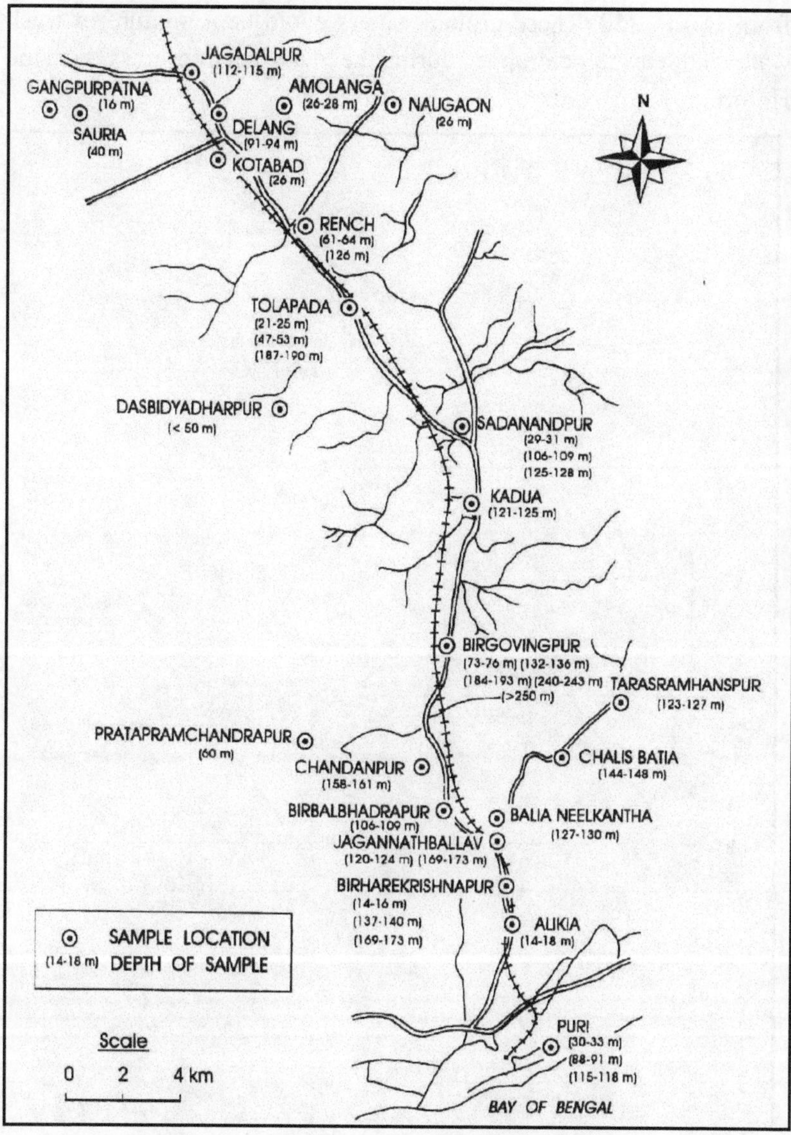

Figure 4.46. Sample locations along the Delang-Puri sector

Radiocarbon ages of groundwater samples from Pratapramchandrapur and Puri are 9850 a BP and 12550 a BP respectively. The EC of the samples varies from 1540 μS/cm at Delang to 20 000 μS/cm at Birgovindpur. Most of the samples are saline. The Cl/Br ratios range from 150 to 250. These ground waters could be a mixture of fresh water and seawater entrapped during the Flandrian transgression in the Holocene.

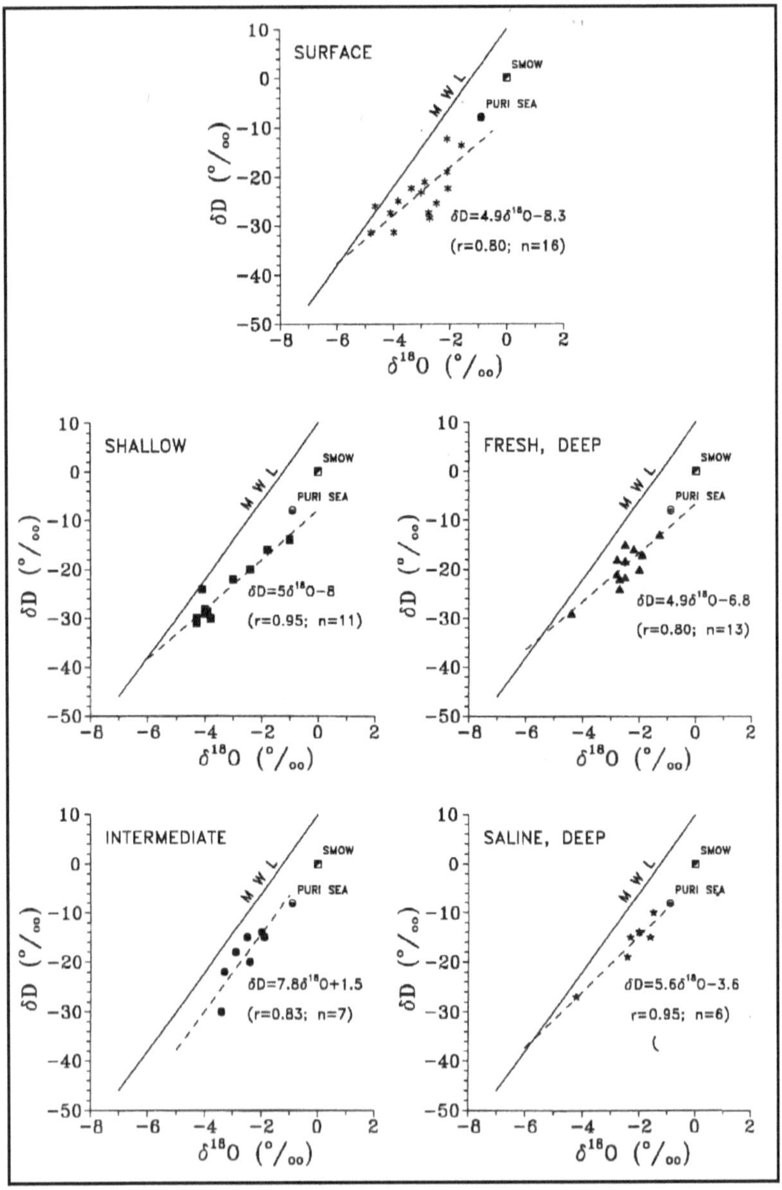

Figure 4.47 δD - δ^{18}O plots of Delang – Puri waters

The groundwater samples in the fresh, deep aquifer (B_2, B_3 in Fig.4.45) fall on a regression line $\delta D = 4.9\, \delta\, ^{18}O - 6.8$ which is similar to the one for surface waters. These samples show high tritium content, 3 –8 TU with an average EC value of 1150 µS/cm. The Cl/Br ratio ranges from 70 to 100. The recharge to these aquifers appears to be relatively fast and could be from outcrops of basement rocks in the northwest and western parts. Old marine waters, if any in these aquifers must have been flushed out by modern recharge from inland sources.

The saline groundwater samples from the deep aquifers (C_1, C_2 in Fig 4.45) fall also on an evaporation line $\delta D = 5.6\, \delta^{18}O - 3.6$. The radiocarbon ages for these samples range from 16,500 a BP to 30,000 a BP. EC ranges from 2430 to 29,700 µS/cm. Cl/Br ratios range from 190 to 340. These waters could be a mixture of fresh and marine waters entrapped during rise in eustatic level during the Late Pleistocene interglacial stage.

The ^{14}C ages are well supported by the radiocarbon age of 27000 a B.P of gastropod and lamellibranch fossils from sediment at a depth of 125 m.

Fig. 4.48 shows ^{14}C versus depth. Groundwater stratification is clearly seen, especially in the deep and intermediate saline aquifers.

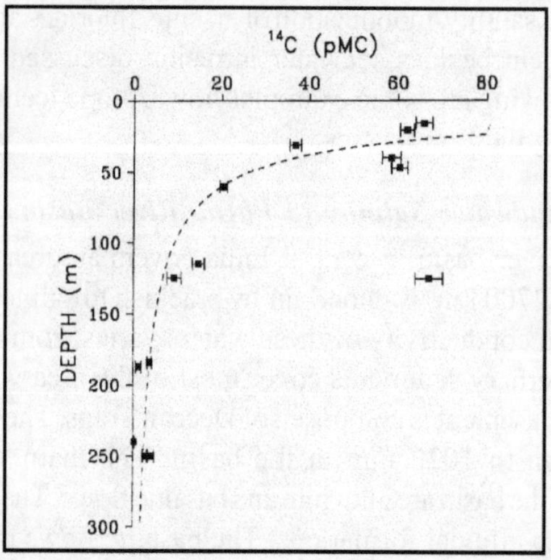

Figure 4.48 Carbon-14 with depth in Delang-Puri aquifers

Conclusions

1. The groundwater salinity in the intermediate aquifer is largely due to the Flandrian transgression in the Holocene
2. Fresh, young and deep groundwater forms a well-developed aquifer at depths of 102 – 165 m, which receives recharge through basement rocks. This could form the main source for groundwater exploitation.
3. The salinity in deep ground waters is due to marine waters entrapped due to rise in eustatic level during the Late Pleistocene interglacial stage.
4. The groundwater in aquifers, particularly the saline aquifers is stratified. The yellow-brown unfossiliferous sediment horizon sandwiched between dark grey fossiliferous horizons probably marks the climatic change, the intertropical aridity during the Late Pleistocene

4.2.5 Groundwater Contamination

Surface water resources are normally contaminated by human activities resulting in industrial and domestic effluents. The sources of these effluents are often easily recognized and controlled. The groundwater systems, on the other hand are subject to contamination due to natural causes as well as due to human activities. Natural causes include inland salinity, mobilization of arsenic, fluorides and nitrate from the environment besides seawater intrusion discussed in the earlier section. Following are some examples how isotope techniques help in characterizing these processes.

4.2.5.1 Groundwater Salinity in Purna River Basin

Purna River basin in central India covers an area of 7500 km^2, out of which 2700 km^2 is underlain by brackish to saline groundwater. The electrical conductivity of these waters varies from 400 to 30,000 μS/cm. Quaternary sediments cover most of the area while the basin margins and basement is composed of Deccan Traps. The rainfall varies from 620 mm to 1020 mm in the basin. The main water-bearing formations in the basin are alluvium and basaltic flows. The inland salinity is restricted to alluvial formations. The basaltic flows generally yield fresh groundwater.

An earlier study had suggested intrusion of ancient seawater from far away Bay of Bengal through Son-Narmada lineament. This was not conclusive.

An isotope geochemical study [30] was carried out to more conclusively investigate the origin of salinity of Purna groundwater. Two sets of water samples (pre-monsoon and post-monsoon) were collected from surface waters, shallow (<15 m) aquifers and deep (> 15 m) aquifers and analyzed for chemistry and isotopes (^{2}H or D, ^{18}O, ^{13}C, ^{34}S, ^{3}H and ^{14}C).

Chemistry

The Piper trilinear plots for fresh waters and saline waters showed that fresh waters are generally Na-HCO_3 type whereas saline samples are predominantly Na-Cl type. The hydrochemical facies in saline waters change from HCO_3 to Cl type. No linear trend in salinity and depth were observed. It is observed that wherever groundwater flow is less, flushing is less and salinity is more. Computation of molar ionic ratios of the samples suggested that the salts have been derived as a result of weathering of the basalt present in the vicinity. Contribution of local precipitation to groundwater as a direct recharge is insignificant except in hard rock area. The foothills of the Satpura range, which is situated in the north of the basin, act as a recharge zone for fresh groundwaters in the area. Here the alluvium cover is quite thick and it becomes thinner and thinner as it progresses towards Purna River. Ultimately, the alluvium acts as discharge zone along the depression of Purna.

Isotopes

The deep aquifers of the area have saline, brackish and fresh waters. The δD-$\delta^{18}O$ plot indicates evaporative enrichment of saline waters. The fresh waters fall near GMWL with a slope of about 8. Brackish waters, falling between saline and fresh waters seems to be mixture of the two waters. This is further inferred as well by the ^{3}H values of the waters. (Fig. 4.49)

Carbon-14 measurements on a few deep saline water samples yielded uncorrected ages of 4100 to 7300 a BP.

In the $\delta^{18}O$ – Cl^- plot (Fig. 4.50) of the deep saline waters, the $\delta^{18}O$ value remains largely unchanged with increase in salinity indicating that these waters derived their salinity by leaching salts from the aquifer matrix, in addition to evaporation.

Final evidence that the salinity could not be of ancient marine origin was obtained by $\delta^{34}S$ measurements on aqueous sulphates. The values ranged from +12.3 to +13.7‰. It is well known that the $\delta^{34}S$ value of marine sulphates remained constant at ~ +20‰ for millions of years [31].

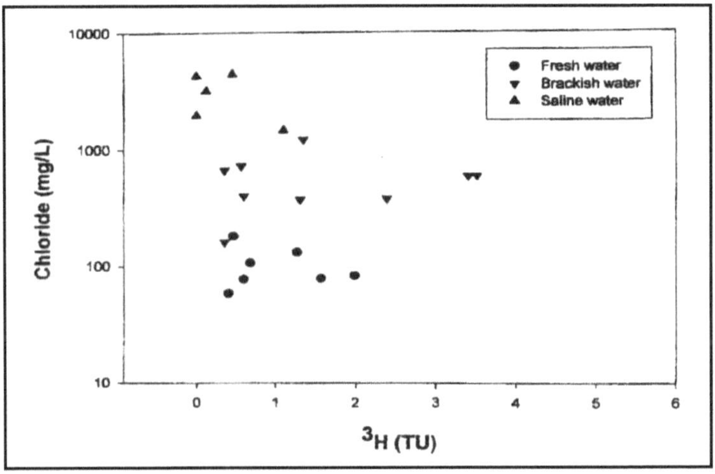

Figure 4.49 Cl^-- ^3H plot of Purna deep aquifer waters (Figure is not in [26])

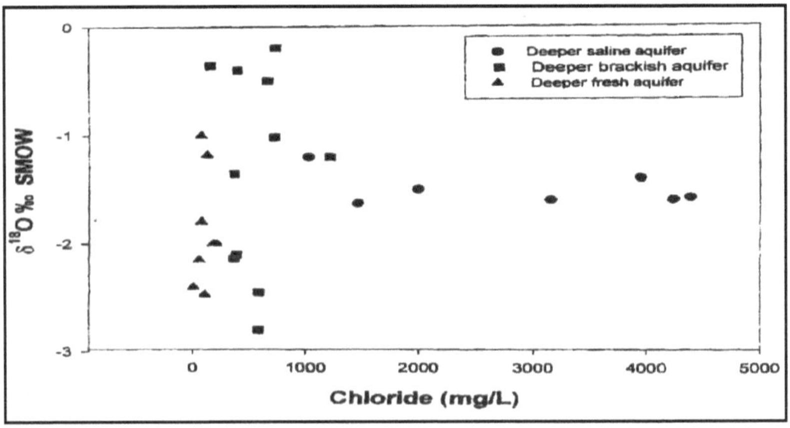

Figure 4.50. $\delta^{18}O$ – Cl^- plot of Purna deep aquifer waters (Figure not in [26])

Conclusion

From the available isotopic and geochemical data of the samples collected in the study, the salinisation mechanism of groundwater in the Purna basin appears to be evaporation of recharging waters coupled with dissolution and leaching of salts from the aquifer matrix and not due to ingression of the past sea.

4.2.5.2 Arsenic Problem in West Bengal – Role of Isotope Geochemistry

Natural arsenic contamination of groundwater in the Bengal basin in Eastern India and Bangladesh is a calamity of incomparable proportions. Over a million people are exposed to this menace in an area of 37000 km^2 covering six districts of West Bengal alone. They are Mushirabad, Nadia, Malda, Barddaman, and North & South 24 Paraganas districts. It is reported that over 200,000 people are already affected by arsenic poisoning. The arsenic levels detected in Bengal groundwater range up to 2.8 mg/L whereas the permissible level in dinking water is 0.01 mg/L, though the commonly quoted value is 0.05mg/L.

Isotope geochemical study in West Bengal

Fig. 4.51 shows the arsenic affected groundwater locations in West Bengal. It is seen that generally all such locations are east of the Bhagirathi River.

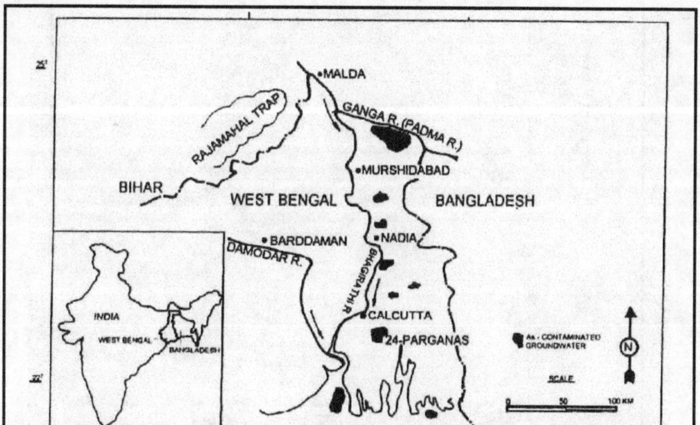

Figure 4.51 Locations of arsenic contamination of groundwater in West Bengal

In West Bengal, arsenic infested aquifers are covered by unconsolidated allocthonous sediment. Interaction of arsenic-bearing sediment with groundwater and the key processes leading to arsenic mobilization into groundwater are important areas of interest. An isotope geochemical study was initiated towards these objectives [32].

In the Bengal basin, the upper delta is covered by Quaternary sediment, primarily formed by alluvial sediment transported by rivers from Chota-Nagpur upland in the west as well as the Himalayan Rivers. Arsenic contamination of groundwater occurs in this sediment and extends eastwards to Bangladesh. The lower delta consists of tidal mud, distributor levees and interdistributory marshy complex formed under a fluvial estuarine and marine environment under the influence of fluctuating sea level in the Tertiary and Quaternary. This plain is characterized by the presence of extensive clay blanket, which is underlain by silt, sand and gravel.

Most of the arsenic affected areas are located in the upper deltaic plain and southeastern part of the lower delta. Arsenic concentration is generally high in fine sand and clay, higher in clay (0.3 to 1.8 mg/L) than in fine sand (<0.2 mg/L). General depth-wise distribution of arsenic concentration in groundwater in several districts is given in Fig. 4.52.

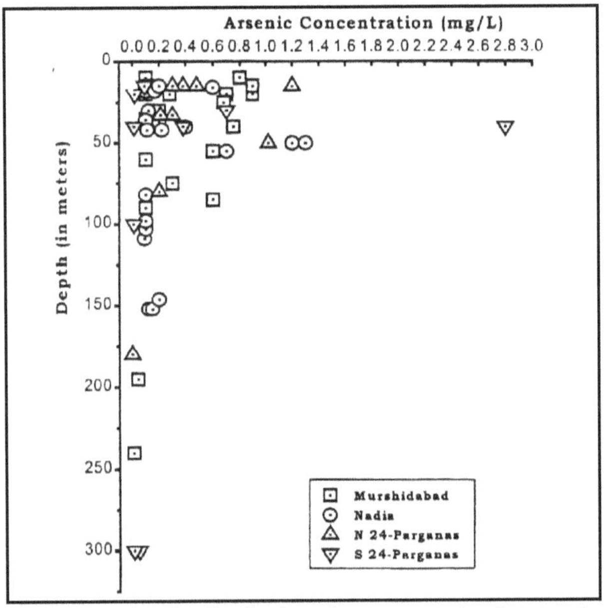

Figure 4.52 Depth-wise distribution of as concentration in groundwater

Hydrochemistry

High concentration of arsenic is generally seen between 10 to 150 m below the ground surface. In Mushirabad district, high concentration of arsenic (0.01 to 0.9 mg/L) was observed in the depth range of 15 to 80 m below ground level. In surface water and in deeper groundwater, the arsenic concentration is generally within acceptable limits (<0.01 mg/L). Both shallow and deep groundwaters are fresh ($Cl^- = <100$ mg/L) with pH values between 6.9 and 7.4. Iron content is in the range of 0.1 to 7.0 mg/L and dissolved oxygen 1.0 to 5.0 mg/L. These are Ca-Mg-HCO_3 type waters. Hydrochemical facies in these waters change from $HCO_3 \rightarrow HCO_3 + Cl \rightarrow Cl + HCO_3$.

In South 24-Paraganas district, arsenic levels in groundwater are higher (0.01 to 2.8 mg/L). The waters are generally Ca - HCO_3 type except at a few places where they are of Na - HCO_3. The pH values are in the range of 6.6 to 7.5, dissolved oxygen 1.9 to 7.3 mg/L and total iron content 0.8 to 8.1 mg/L.

It is observed that the arsenic content in deep groundwater is generally correlated with bicarbonate and iron and inversely correlated with dissolved oxygen.

Isotope data

A large number of samples from Mushirabad and South 24-Paraganas districts have been analyzed for ^2H and ^{18}O and a few for ^3H, ^{13}C, ^{14}C and ^{34}S.

In Mushirabad, δ^2H - $\delta^{18}O$ composition of both shallow and deep ground waters (Fig. 4.53) indicates similar recharge process. The river waters (Surface water in Fig.) are well depleted as they originate at high altitudes. They may not be contributing significantly to groundwater recharge. The major recharge source appears to be local precipitation.

The stable isotope composition (δ^2H, $\delta^{18}O$) of the shallow, intermediate and deep ground water is similar indicating interconnection. The tritium values of the shallow groundwater are in the range of 2 to 10 TU indicating modern recharge, supported by the carbon-14 levels of 79 to 109 pMC. $\delta^{13}C$ values of a few shallow groundwater samples fall in the range of –6 to 18 ‰ indicating possible oxidation of organic mater in the aquifer.

Deeper groundwater in Mushirabad shows tritium values of <2 TU and carbon-14 values of 78 to 85 pMC indicating the water is relatively older.

In South 24- Paraganas district, the shallow groundwater is slightly depleted in δ^2H and δ^{18}O compared to deeper groundwater. The δ^2H and δ^{18}O values of shallow groundwater are in the range of -37‰ to -22‰ and –6‰ to –3.5‰ respectively whereas the values for deep groundwater are in the range of –9‰ to –16‰ and –6‰ to –3‰ respectively. It is thus evident that the sallow and deep ground waters are not interconnected. The tritium values of the shallow groundwater are between 1 and 2TU whereas the deep groundwater values are <1 TU.Carbon-14 levels of shallow groundwater are about 85 pMC whereas those of deep groundwater are about 20 pMC. The carbon-14 ages (using Pearson Model, Eq. 2.24) work out be in the range of 5000 to 13000 a B.P.

Here again, the δ^{13}C values of a few shallow groundwater samples fall in the range of –3.2‰ to –4.1‰ indicating possible inorganic oxidation in the aquifer system.

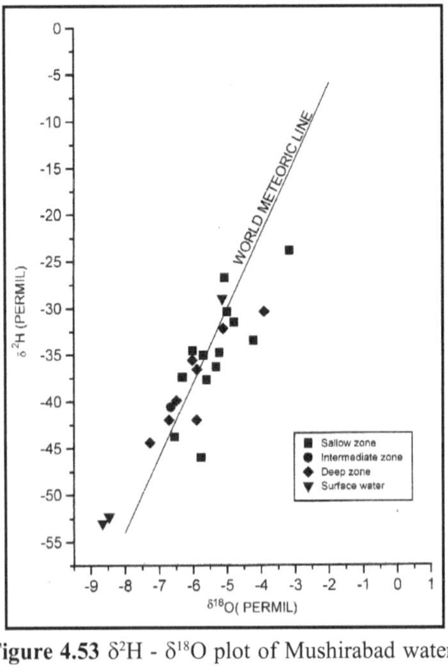

Figure 4.53 δ^2H - δ^{18}O plot of Mushirabad waters

Discussion and Conclusion

Isotope data show that in Mushirabad, the shallow and deep groundwaters are interconnected whereas in the South 24-Paraganas district, they do not appear to be interconnected. This indicates that in Mushirabad, the deeper waters are vulnerable to arsenic contamination from the shallow region whereas in the South 24-Paraganas such a possibility is remote. Hence greater care needs to be exercised while pumping arsenic-free water from the deeper aquifer in Mushirabad.

It is suggested, based on hydrochemistry, that the elevated arsenic concentration in shallow groundwater is possibly due to its release from younger alluvium when arsenic-rich iron oxyhydroxides are reduced in anoxic environment in the aquifer.

The International Atomic Energy Agency, Vienna carried out a similar isotope study [33] on Bangladesh groundwater infested with arsenic and the observations are broadly similar to those of the West Bengal Study. Subsequently [34], the authors recommend isotopic composition as a better criterion on the safety of exploitation than the depth alone. Inorganic oxidation leading to enrichment of ^{13}C is ruled out as a possible mechanism of arsenic mobilization. No specific process of mobilization could be proposed based on hydrochemistry and $d^{13}C$ data. Comparing the isotope data collected earlier in 1979 with the data collected in 1999, it was concluded that the exponential increase in groundwater exploitation in the intervening period had no effect on the arsenic mobilization process.

Another study [35] conducted by a team of institutions led by the British Geological Survey came to the same conclusion based on the depth-wise tritium profile in Bangladesh wells. At a depth of ~40 m in each profile where the arsenics concentration was ~200mg/L, the tritium concentration was ~0.5 TU representing water from the pre-thermonuclear era (>50 years). This shows that arsenic mobilization existed before the general high-level exploitation of groundwater.

4.2.5.3 Nitrate Pollution in Makutupora Basin, Tanzania

The aquifers in the Makutupora basin in central Tanzania (Fig.4.54) are being pumped at a rate of 24000 m^3 /day through a well field of high-yielding boreholes (~500 m^3 /h) drilled along a fault line to supply water to the capital Dodoma. When the extraction rates sharply rose in the 90's, nitrate levels rose 3 times above the WHO limit of 45 mg/L for drinking water. By late 90's when the extraction rates were reduced to normal levels, the nitrate levels returned to those observed in the early 90's. The increase in the nitrate level coincided with the failure of a grape production scheme in the area and the farmers had to plough up vineyards for subsistence farming. This gave rise to the suspicion that agriculture was responsible for nitrate pollution and plans were made to move several villages from the catchment of the basin and stop agriculture [35].

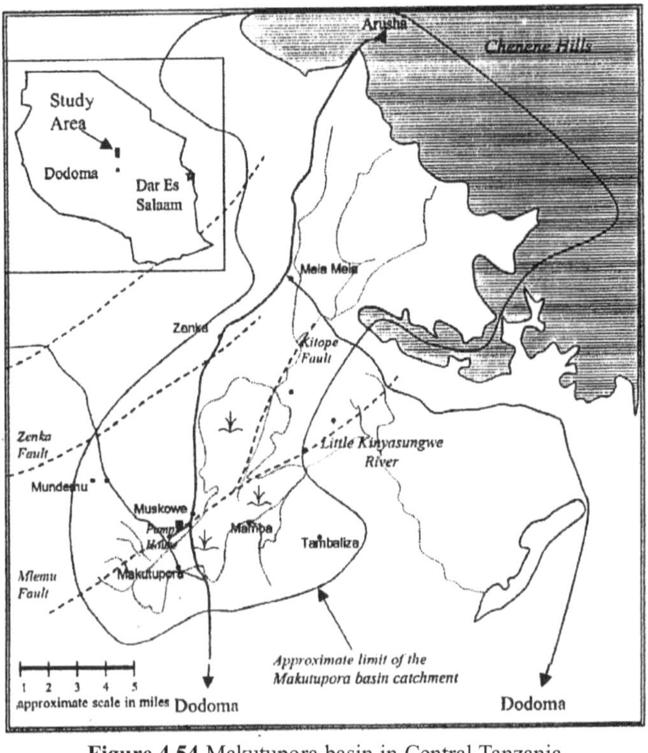

Figure 4.54 Makutupora basin in Central Tanzania

But a simple calculation showed that about 3 tons/day of nitrate would be needed to produce the observed level ~100 mg/L at the peak extraction rate of 30000 m^3 /day. As such level of nitrate input did not exist, alternate possibilities were investigated.

The deep aquifer in the basin in weathered granite is confined capped by a clayey layer. An isotope study [35] during 1999-2002 found no measurable tritium implying a residence time of over 200 years. This showed that it was impossible for the nitrate to increase due to human activities of 1 – 2 years. Carbon-14 value of 60 pMC corresponding to an upper limit of 3000 years of residence time further strengthens the above conclusion.

The confined aquifer is probably recharged from the hills to the north of the well field and leaks upward into the shallow aquifer under natural pressure condition.

The only explanation for the high nitrate condition is that, under the influence of the intense exploitation in the 90's, the pressure gradient in the vicinity of the wells was reversed and shallow groundwater (presumably high in natural nitrate) was being drawn in (see Fig. 4.55). Once the pumping rate was reduced, the upward pressure gradient was restored. It is obvious that high nitrate concentration was historically present in shallow groundwater, mainly from non-anthropogenic sources, but masked by the upward leakage from the confined aquifer.

Isotope data certainly sounded a word of caution on proposed remedies involving moving populations and suggested control on pumping rates.

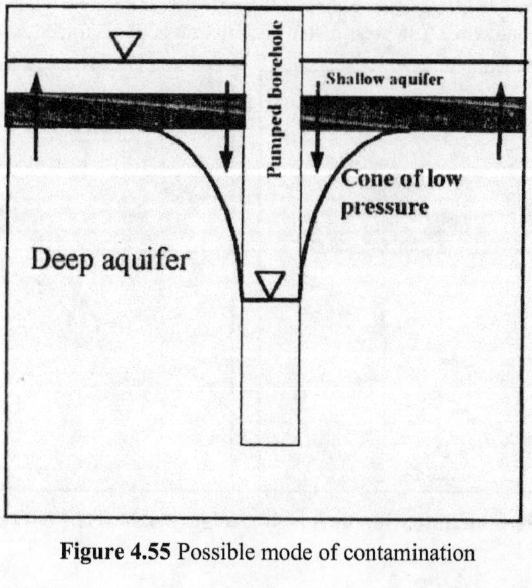

Figure 4.55 Possible mode of contamination

4.2.5.4 Source of pollution of groundwater at an industrial site [36]

Groundwater in the premises of a Rare Earths plant in Cochin (Kochi), Kerala showed contamination. Chemical data indicated that the water was highly acidic with pH in the range of 3 to 7, had high sulphate (up to 1100 mg/L), nitrate (up to 450 mg/L), phosphate (up to 10 mg/L) and fluoride (up to 6 mg/L). Fig.4.56 shows the location of the study area and Fig. 4.57 the contamination levels. The plant is situated west of the Periyar River.

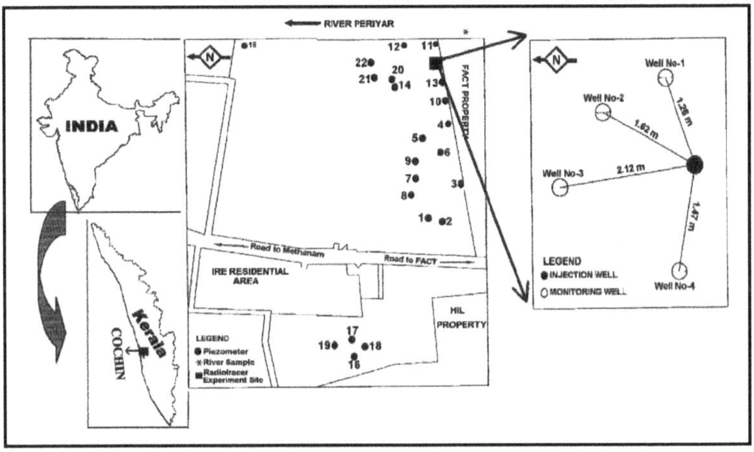

Figure 4.56 Study area (9• piezometer no., Box on Right is tracer experiment area)

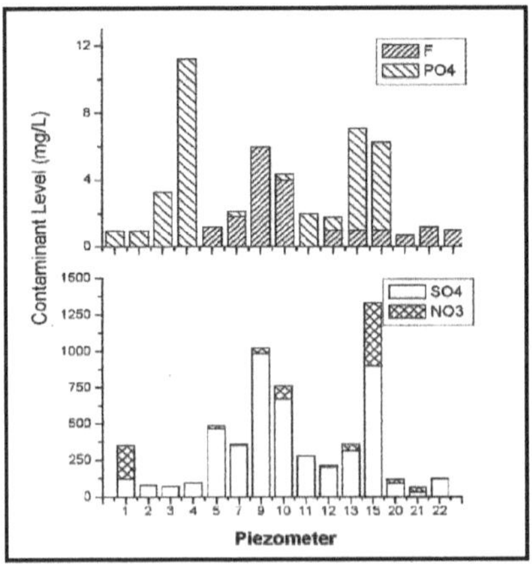

Figure 4.57 contamination levels in piezometers

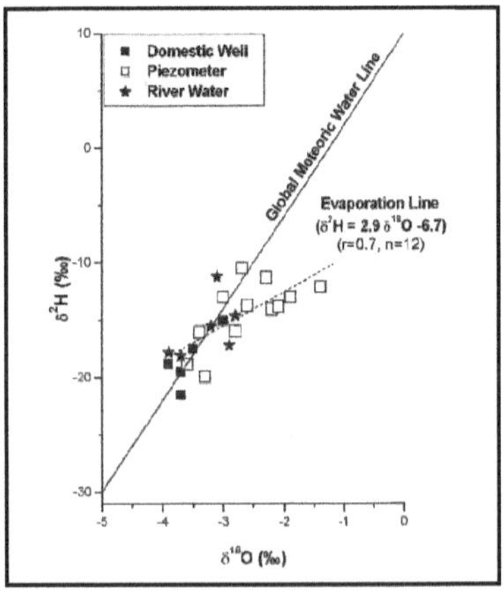

Figure 4.58 δ^2H - $\delta^{18}O$ plot

An isotope investigation involving environmental stable isotopes as well as injected radiotracers was carried out [36] to characterize the pollutant transport and identify the source of pollution.

Water samples from the piezometers, Periyar River and other uncontaminated wells in the plant site were analyzed for ^2H and ^{18}O. The δ^2H - $\delta^{18}O$ plot shown in Fig. 4.58 indicated that many contaminated samples from the piezometers fall on an evaporation line indicating that the contaminated waters were subjected to evaporation prior to infiltration.

The velocity and direction of transport of the contaminated waters were determined by two experiments using injected radiotracers. One was to determine the filtration velocity using a single borehole and the second is to determine the interstitial velocity and direction of groundwater flow using a multi well method (see 2.2.13 for the description of the methods).

Determination of filtration velocity

Here is some element of innovation in the method adopted. Instead of isolating a part of the borehole, the tracer was injected directly into the borehole and the entire column of water was treated as the dilution volume and well mixed as and when a tracer log was taken. About 7.5 MBq (200 μCi) of ^{82}Br as NH_4 Br was injected into a borehole and was well mixed in the borehole column. The column was logged 3 times using a waterproof NaI scintillation counter connected to a count rate meter. The logs are presented in Fig. 4.59. Integrating each log and calculating the decrease in the total count rate after each time interval, the rate of dilution was computed. Filtration velocity was then determined as 1.3 cm/day using Eq.2.12.

Determination of interstitial velocity and direction

The arrangement for the injection well and the detection wells for the multi-well experiment is shown in Fig.4.56. About 40 MBq (1 mCi) of ^3H as tritiated water was injected into the injection well and samples were taken at regular intervals for tritium assay in a liquid scintillation counter. The tritium concentration data is presented in Fig. 4.60

From the tracer data, it is seen that the direction of movement of polluted waters was towards north and northeast and the flow velocity was 1.2 to 1.4 cm/ay.

From the hydrochemical, sable isotope and radiotracer data, it could be concluded that the source of high sulphate, nitrate, phosphate and fluoride contamination could be south of the plant and possibly from the effluent channel of a nearby chemical factory.

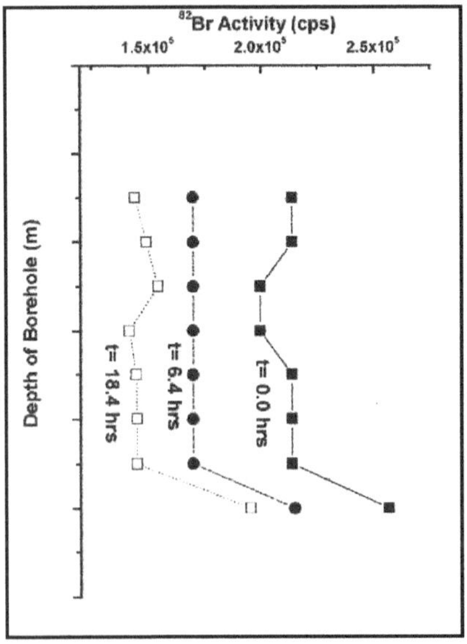

Figure 4.59 ^{82}Br logs in borehole

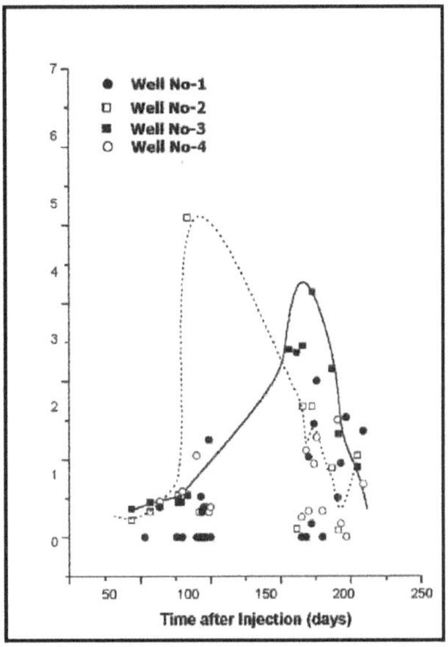

Figure 4.60 Relative tritium tracer level (y-axis) vs. time.

4.2.5.5 Origin of saline ground waters in the Valencia Lake region, Venezuela

Background

The Valencia Lake in Venezuela is situated in an intermountain plain near the city of Caracas close to Caribbean Sea coast. It has a surface area of 360 km^2. The catchment area was 2600 km^2, but was artificially increased to 3200 km^2 by a shortcut to a neighbouring catchment in 1981. This caused an additional input of 5 m^3/s and the lake level rose from 402 m to 406 m a.s.l. Many riverbeds enter the lake (see Fig. 4.61), but there is no surface outflow. The subsurface discharge is, however active as the electrical conductivity is low (~2000 μS/cm)

Today the lake receives untreated effluents from two cities; Maracay and Valencia, as well as from wood processing and chemical plants and also from cattle sheds. It is estimated that about 10 tons of nitrogen and 3 tons of sulphur reach the lake every year. The chemical pollution led to complete eutrophication of the lake and as a result fishing and swimming are forbidden.

The intermountain plain is highly urbanized and agriculture is mostly by lift irrigation (~3000 wells). In addition, the Taguaiguay reservoir in southeast is used for irrigation. The groundwater table got lowered below the lake level. The groundwater quality in the north and east of the lake deteriorated with increase in sulphate levels. Possible causes for this deterioration are:

- Infiltration of Valencia Lake waters
- Upwelling of deep sulphate-rich groundwater into the overlying fresh groundwater
- Irrigation return waters enriched in agrochemicals

An environmental isotope study [37] supported by hydrochemistry helped in identifying the origin of sulphate waters.

Hydrogeology of the area

The catchment area around the intermountain basin is composed of crystalline rock whereas its weathered products cover the plain. The fluvio-lacustrine sediments consist of silt, gavel and sands and in some places gypsum is exposed.

The crystalline rock is of low permeability allowing formation of a dense network of surface streams whereas the weathered products in the plain are so permeable that intense bank infiltration prevents any surface input to the lake. Groundwater extraction is ~ 14 m^3/s.

Isotopes and chemistry

Fig. 4.62 shows the δ^2H - $\delta^{18}O$ plot of the groundwater samples east of the Valencia Lake. The isotope data of groundwaters in the southeastern irrigation area fall on an evaporation line away from the meteoric water line. This must be due to the mixing of groundwater with evaporated waters from the surface reservoirs or with irrigation return waters. Since isotopically enriched waters were found in areas with groundwater table above or below the lake level, infiltration of lake waters could be completely or partially ruled out.

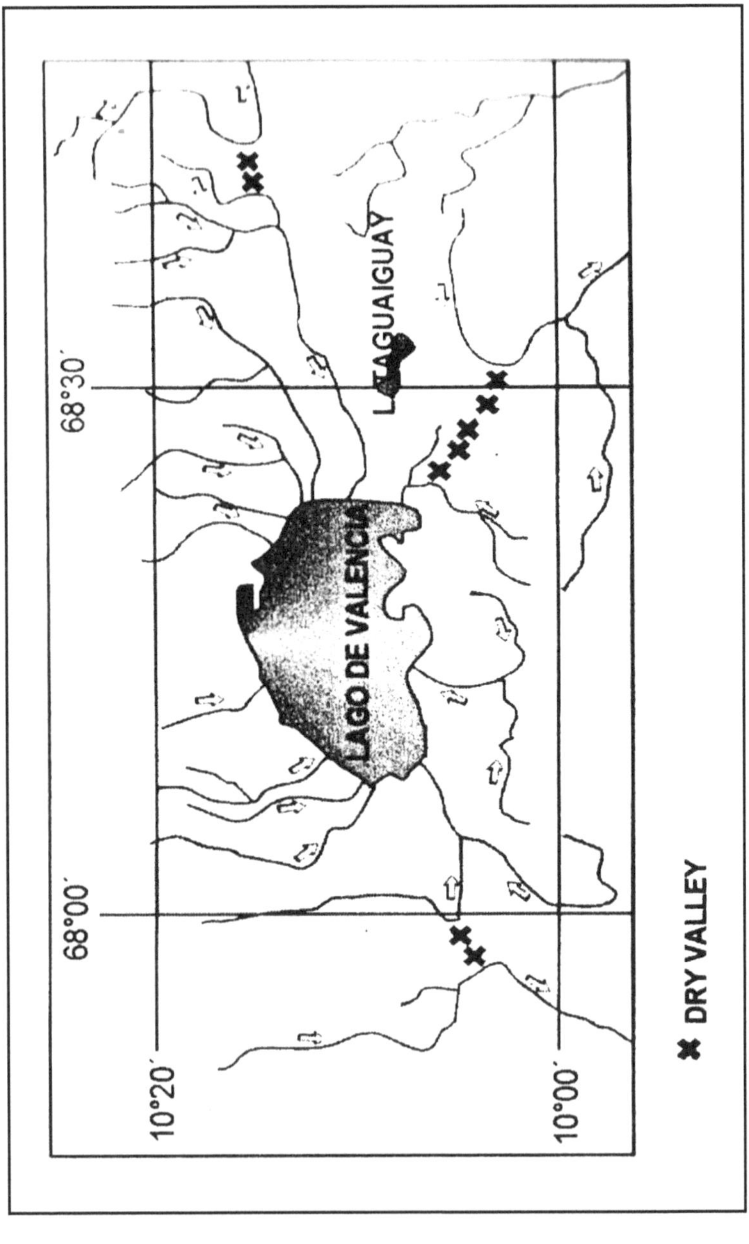

Figure 4.61 Valencia Lake and its catchment area

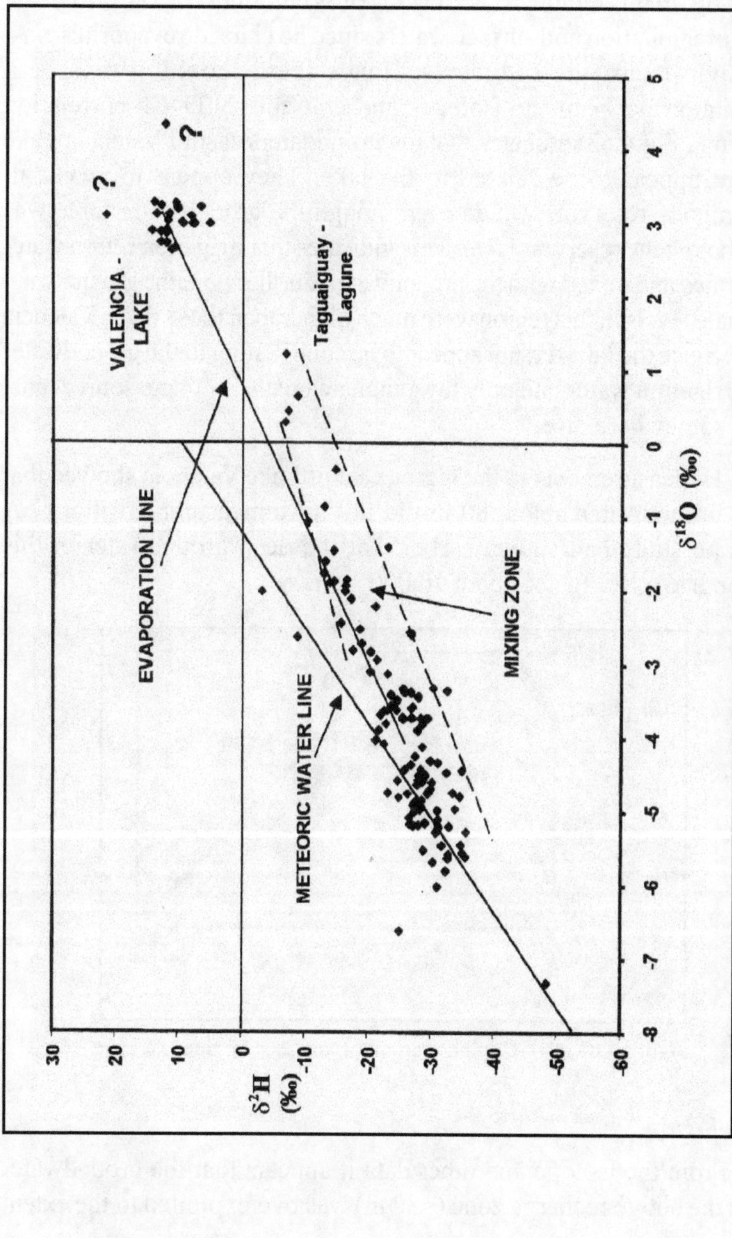

Figure 4.62 δ^2H - $\delta^{18}O$ plot of groundwaters east of Valencia Lake

In the southeast, sulphate (up to 2000 mg/L) and chloride (up to 150 mg/L) were higher than elsewhere and electrical conductivities were even higher than those in the Valencia Lake or Taguaiguay reservoir. In this region, ^{2}H & ^{18}O as well as chlorides can only originate from precipitation and surface waters since no chloride evaporites exist in fluvio-lacustrine sediments. Hence there should always be a correlation between the isotopes and chloride. $\delta^{18}O$–Cl^- correlation shown in Fig.4.63 indicates that the groundaters east of Valencia Lake do not appear to be related to the lake. They appear to mix with Taguaiguay reservoir waters even though the groundwater table was well below the reservoir level. This indicates that irrigation return water infiltrated and mixed with the groundwater. Furthermore the groundwater sulphate levels in this region were much higher than those in the Valencia Lake. Since the lake did not appear to be contributing to the groundwater and irrigation water had only low sulphate levels, sulphate source must exist somewhere else.

^{3}H measurements in the region east of Lake Valencia showed that deep groundwater below 80 m did not have measurable tritium. A separate study had shown earlier that the deep groundwater in this region had a ^{14}C age of about 10,000 years.

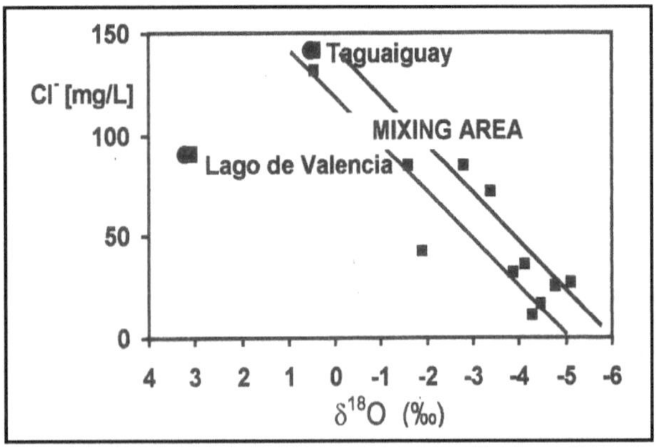

Figure 4.63 $\delta^{18}O$–Cl^- plot of groundwaters east of Valencia Lake

Conclusion

From the isotope and other data it appears that the groundwater from the active recharge zone (<80 m) was overexploited to the extent the older groundwater (high in sulphates due to leaching of gypsum) from the deeper and passive recharge zone intruded upwards through forced hydraulic shortcuts and caused deterioration of the groundwater quality in the active recharge zone. Fig. 4.64 shows the inferred scenario.

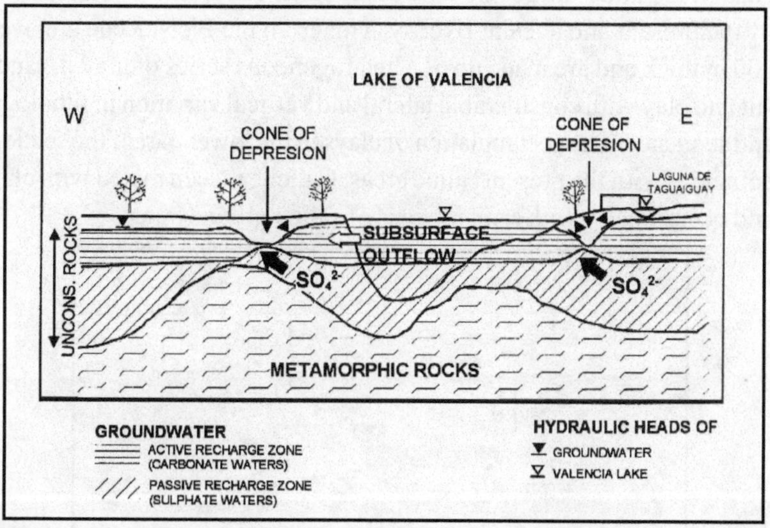

Figure 4.64 Probable scenario of sulphate water intrusion into active recharge zone

4.2.5.6 Palaeo-salinitiy of Haryana groundwaters

Haryana state in northwest India is about 44000 km^2 in extent and lies astride a low watershed between the Indus and Ganga River systems. River Yamuna, which is a tributary of Ganga, forms the eastern boundary of the state whereas the River Ghaggar, which belongs to the Indus system, flows along the state's northern boundary. In addition to these two river basins, there is an internal drainage basin without a drainage outlet. Over 60% of the area is underlain by deep (>50 m) saline aquifers and about 10% and 50% has also shallow saline and brackish aquifers respectively. The climatic zones are wet and subhumid in the northeast and arid and semi arid in the southeast. Fresh groundwater mainly occurs in the northern part of Haryana whereas most of the saline areas are in the southern part of the sate. The rise in the saline water table due to seepages from the canal network is an additional cause of anxiety. In view of this an isotope-geochemical investigation was carried out [38] to obtain information on the cause of salinity and to plan remedial measures.

Hydrogeology

Fig. 4.65 shows the geological map of Haryana and Fig. 4.66 shows a generalized geological section NE – SW across the state. The Quaternary alluvium covers 97% of the state. These deposits laid down by the present and ancient river systems from the Pleistocene are over 300 m thick and are made up of a heterogeneous series of gravel, sand, silt and clay with considerable lateral and vertical variation in lithology and sequence. The accumulation of clays in the lower part of the section coincides with the present saline areas. Caliche is seen mixed with clay and occasionally with sand in areas of saline groundwater.

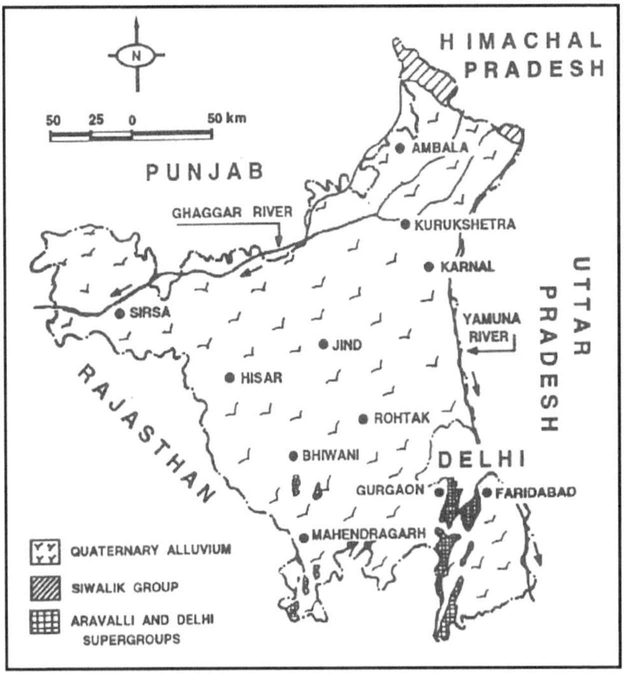

Figure 4.65 Geological map of Haryana

Hydrochemistry

The groundwater in the study area (Fig. 4.67) mainly occurs under water table conditions though semi-confined to confined conditions exist in some places. Perched water table is seen in the sand dunes near saline areas. The area is semi-arid with an annual rainfall of about 200 mm.

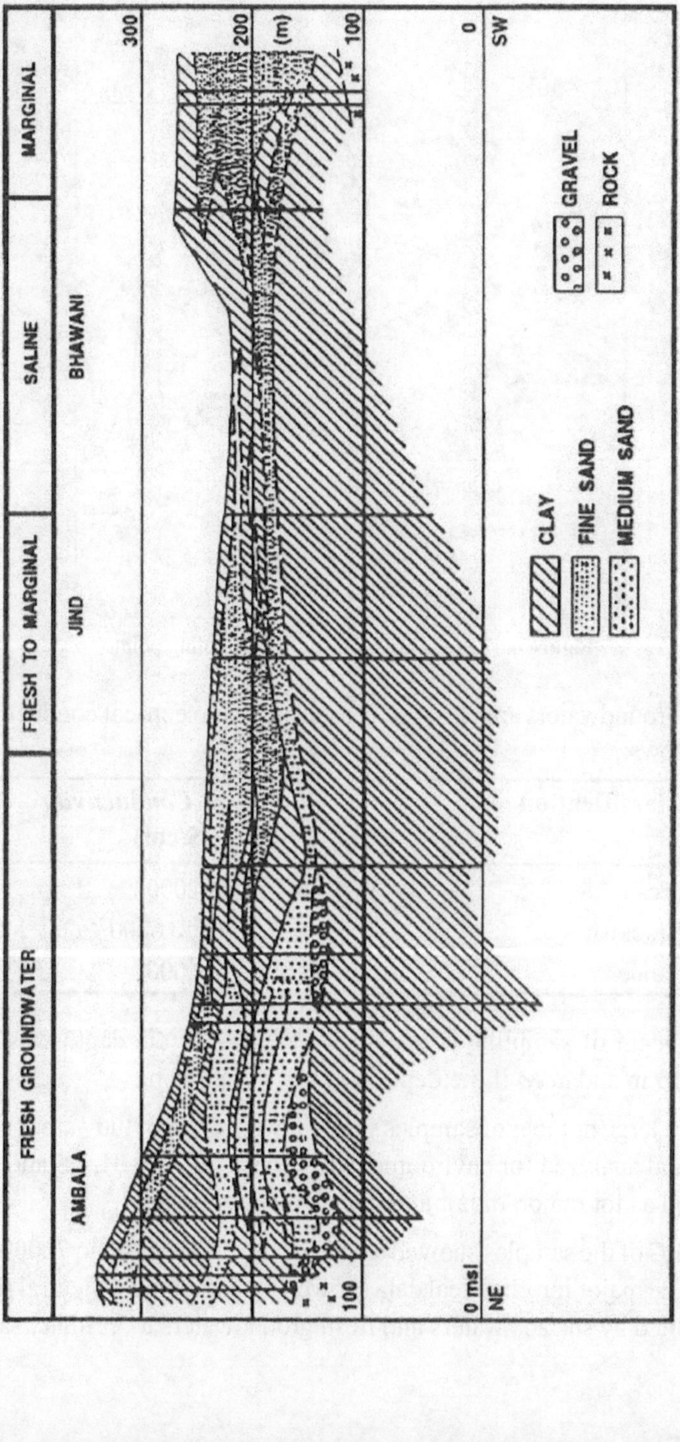

Figure 4.66 Generalized geological cross-section NE-SW across the state

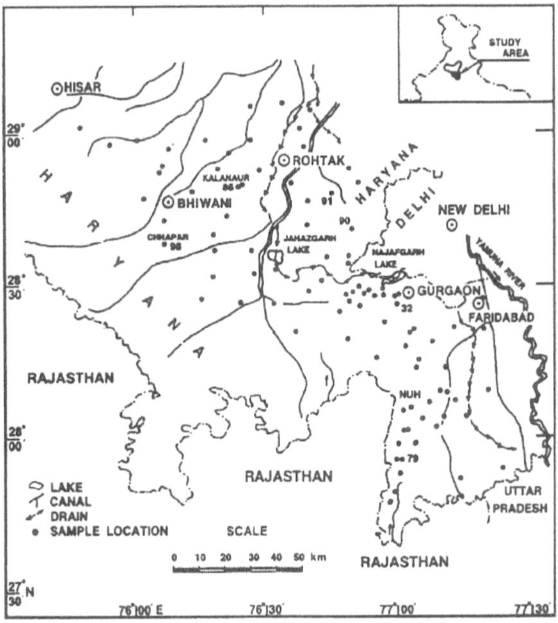

Figure 4.67 Study area showing the sampling points

Groundwaters are classified based on their electrical conductivity as follows:

Classification	***Electrical Conductivity* (µS/cm)**
Fresh	<2000
Brackish	2000 to 6000
Saline	>6000

Depth of sampling is denoted as *shallow* if the depth was less than 15 m and *deep* if the depth was more than 15 m.

A large number of samples were drawn from the study area (Fig.4 .67) and analyzed for environmental isotopes (D, ^{18}O, ^{3}H, ^{34}S and ^{14}C) as well as for major, minor and trace ion chemistry.

EC of the samples showed a wide variation from 300 to 72000 µS/ cm. The major ion chemical data showed transition from 'neutral type' exhibited by surface waters and fresh groundwaters to 'primary saline type' exhibited by saline groundwaters indicating that the salinity could have been derived from the surrounding alluvium.

In most of the saline waters, sodium was the dominant cation with subordinate calcium and magnesium and chloride was the dominant anion with subordinate sulphate. In general EC and chemical type vary as follows:

EC (μS/cm)	Chemical type
<2000	$HCO_3 > Cl > SO_4$
2000 – 6000	$HCO_3 > SO_4 > Cl$
>6000	$Cl > SO_4 > HCO_3$

Three distinct areas had chloide levels above 10000 mg/L (see Fig. 4.56). These were Gurgaon, Nuh and Kalanur (Fig. 4.68). These are depression areas, playas and/or interdunal depressions.

The strontium content of saline waters was high (up to 70 mg/L). This could be due to incongruent dissolution of aragonite. This is a slow process and indicative of long contact with aquifer matrix. The iodide content in saline groundwater is also considerably high (up to 0.79 mg/L). Decomposition of organic matter and release of iodide is the possible process.

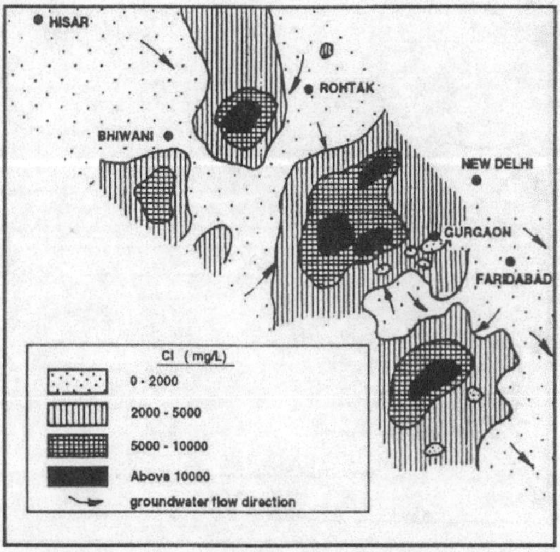

Figure 4.68 Chloride distribution in groundwaters in the study area

In general, the saline groundwaters exhibit chemical enrichment due to high evaporation (in depression areas without drainage outlet) and dissolution of salts from the matrix during long residence time in the aquifers.

Isotopes (D & ^{18}O)

New Delhi was one of the stations of the IAEA/WMO network for data on environmental isotopes in precipitation [39], operational since 1960. Due to the proximity of this station to the study area, the following local meteoric line for Delhi could be considered for the study area also.

$$\delta = 8\ \delta^{18}O + 8.4$$

This line is close to the Global Meteoric Water Line.

The stable isotope compositions of the irrigation waters derived from Ghaggar and Yamuna Rivers were as follows:

River	δD (‰)	$\delta^{18}O$ (‰)
Ghaggar	-62.9	-9.71
Yamuna	-48.4	-7.95

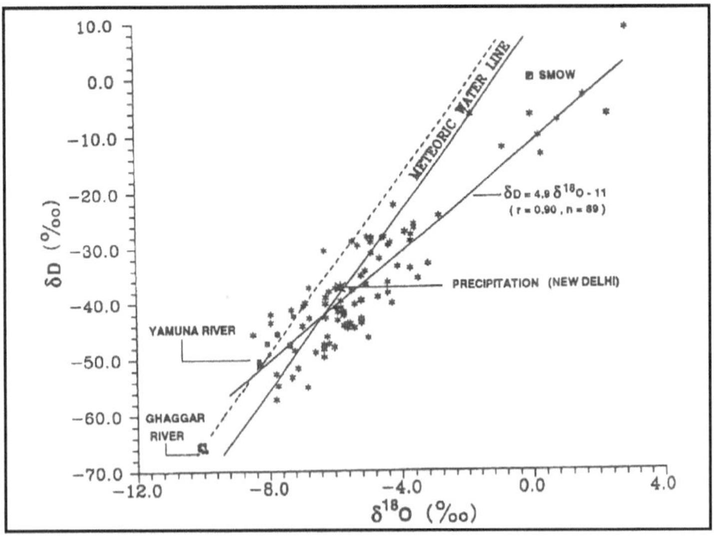

Figure 4.69 δD - $\delta\ ^{18}O$ plot of Haryana groundwaters

The stable isotope composition of wells sampled in the study area showed wide variation with $\delta^{18}O$ values varying from –8.5 to +2.9‰. As can be seen in Fig. 4.69 a number of samples showed more depleted stable isotope composition compared to the precipitation in New Delhi indicating the significant contribution from irrigation return waters to groundwater replenishment. The surface water samples plot well above the MWL (dotted line above WML) and are represented by the equation: $\delta = 7.95\ \delta^{18}O + 14.5$

This indicated that the hydrometeorological conditions in the catchments of Ghaggar and Yamuna Rivers were different from New Delhi.

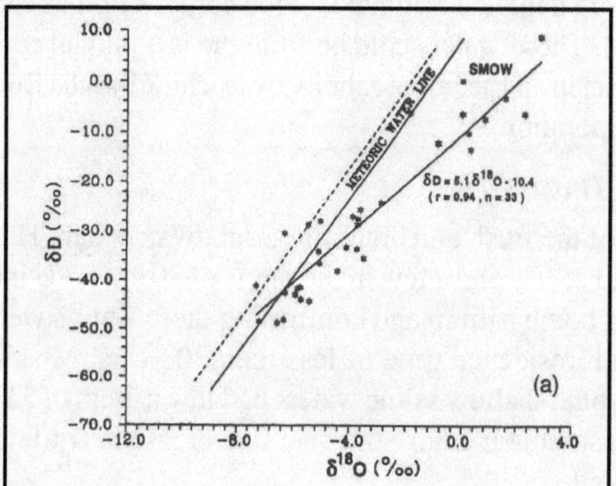

Figure 4.70 δD - $\delta^{18}O$ plots of (a) shallow saline and (b) deep saline groundwaters

The groundwater samples in the study area fall on an evaporation line with slope 4.9 in the δD - δ ^{18}O plot in Fig. 4.68 suggesting evaporation as the main salinization process.

Some of the shallow saline waters had highly enriched stable isotope composition as shown in Fig. 4.70(a). It is interesting to note these samples represent areas where salt extraction was active during 19th century. These waters thus show high evaporation as a consequence of the saltpan activity (see similar observation in Minjur aquifer study in 4.2.4.1).

Most of the deep saline waters also show some evaporation effect, but with more depleted original isotope composition as can be seen in Fig. 4.70(b). These waters cold be from the last pluvial episode before aridity had set in and acquired salinity by leaching of salts from sediments besides evaporation.

Isotopes (^{3}H & ^{14}C)

Most of the fresh and brackish groundwaters had ^{3}H levels in the range of 5 to 25 TU and a few in the range of 30 to 45 TU indicating the presence of bomb tritium and confirming these waters were relatively modern with residence time of less than 30 years. Most of the deep saline and some shallow saline waters had low tritium of 5TU and some had no measurable tritium. In general decrease in tritium with depth was observed.

The following table gives ^{14}C and other data for some selected deep saline groundwater samples.

Place	Depth (m)	EC (μS/cm)	^{3}H (TU)	^{14}C (pMC)	Uncorrected Age (a BP)
Harasu	20	17000	0.9	30.9	9700
Naseerbas	63	8150	0.9	72.5	2700
Kalanaur	27	71700	n.d	64.4	3600
Dadoba	20	51100	n.d	61.2	4100
Sampla	20	68100	n.d	94.0	500
Chappar	62 - 84	30000	1.3	35.9	8500

From the above table, it appears that the deep saline groundwaters were formed during the Holocene confirming D and ^{18}O data in Fig. 4.70(b).

Conclusion

The isotope data along with chemistry show that the shallow brackish waters are largely modern and owe their salinity to evaporation of stagnant waters in the depression areas. The saline waters, on the other hand, are old and were formed during the Holocene. Their salinity could be partly due to evaporation and partly due to leaching of salts left behind by the earlier arid phase. Amelioration of the climatic condition might have helped to flush out the saline waters from most of the shallow and some deep aquifers. But, some shallow and most deep aquifers have retained their saline waters.

References

1. K.O. Munnich (1983), Moisture movement in the unsaturated zone, In: Guidebook on Nuclear Techniques in Hydrology, Technical Reports Series No. 91, IAEA, Vienna pp. 203 – 222
2. S.S.D. Foster, A. Smith-Carrington (1980), The interpretation of tritium in the chalk unsaturated zone, J. Hydrol., 46, pp.343 – 364
3. B.S. Sukhija, Rama (1973), Evaluation of groundwater recharge in semi-arid region of India using environmental tritium, Proc. Indian Acad. Sci., Vol.77, no.6, pp. 279 – 292
4. U. Zimmemann, K.O. Munnich, N. Roether, W. Kreutz, K. Schubach, C. Siegel (1966) Tracers determine movement of soil moisture and evapotranspiration, Science, 152, pp. 346
5. A. Radhakrishnan Nair, A.S. Pendharkar, S.V. Navada, S.M. Rao (1979), Groundwater recharge studies in Maharashtra – Development of isotope techniques and field experience, Isotope Hydrology 1978, Vol.II, Proc. A Symposium, IAEA, Vienna, pp. 803 – 826
6. P.S. Datta, P.S. Goel, Rama, S.P. Sangal, Hans Kumar, Prakash Bahadur, R.K. Sabharwal, B.S. Tanwar (1975), Tritium tracer studies on groundwater recharge in the alluvial deposits of Indo-gangetic plains of western U.P, Punjab and Haryana, Proc. Indo-German Workshop on Approaches and Methodologies for Development of Groundwater Resources, NGRI, Hyderabad, India, pp. 309
7. P.S. Datta, S.K. Gupta, A. Jayasurya (1978), Soil moisture movement through vadose zone in alluvial plains of Sabarmati Basin, Physical Research Laboratory, Ahmedabad, India Report No. HYD-78-03
8. R.N. Athavale, C.S. Murti, R. Chand (1980), Estimation of recharge to phreatic aquifers of the Lower Maner Basin, India using the tritium injection method, J. Hydrol. 45, pp.185

9. R. Rangarajan, R.N. Athavale (2000), Annual replenishable groundwater potential of India – an estimate based on injected tritium studies, J. Hydol. 234, pp. 38 - 53

10. S.M. Rao (1983), Use of radioactive tracers in studies on infiltration through unsaturated zone, In: Tracer Methods in Isotope Hydrology, IAEA-TECDOC-291, IAEA, Vienna, pp. 47 – 66

11. G.S. Tazioli, P.M. Cantori, G.F. Ciancetti, R. Dazzi, G. Gatto, B. Matticchio, G. Mozzi, G. Zmabon (1995) Intercomparison of different tracers in the evaluation of groundwater dynamics in homogeneous porous aquifers – A study in the alluvial plains of Venice, Isotopes in Water Resources Management, Proc. of a Symp., IAEA, Vienna, pp. 147 – 159

12. H. Beherns (1983), "Comparison of radioactive and non-radioactive tracers", Tracer Methods in Isotope Hydrology, IAEA-TECDOC-291, IAEA, Vienna, pp 173-185

13. R.N. Athavale, R. Rangarajan, S.M. Rao (1992), Determination of flow direction of geothermal waters at Manikaran using the borehole tracer technique, Journal Geological Society of India, Vol. 39, PP. 329 – 337

14. S.M. Rao, S.K. Jain, S.V. Navada, A.R. Nair, K. Shivanna (1987), Isotope studies on seawater intrusion and interrelations between water bodies – some field examples, Isotope Techniques in Water Resources Development, Proc. Intl. Symp, IAEA, Vienna, pp. 403 – 425

15. K. Shivanna, U.P. Kulkarni, T.B. Joseph, S.V. Navada (2004), Contribution of storms to groundwater recharge in the semi-arid region of Karnataka, India, Hydrol. Process. 18, pp. 473 – 485

16. I. Fórizs, T. Berecz, J. Deák (1999), Origin and quality of bank filtered groundwater near Budapest: An evaluation of isotope data, Isotope techniques in water Resources Development and Management, IAEA Symposium, Vienna Session 2, pp 27 - 34

17. B.S. Sukhija, D.V. Reddy, P. Nagabhushanam, Syed Hussain, V.Y. Giri, D.J. Patil (1996), Environmental and injected tracers methodology to estimate direct precipitation recharge to a confined aquifer, Journal of Hydrology, 177, pp. 77 – 97

18. R. Rangarajan, S.D. Deshmukh, D. Muralidharan, T. Gangadhara Rao (1989), Recharge estimation in Neyveli groundwater basin by tritium tagging method, Tech. Rep. No. NGRI-89-ENVIRON-65, pp.1 – 63

19. Ninad Bodhankar, S. Upadhyaya, K.M. Kulkarni, S.V. Navada, U.K. Sinha, U.P. Kulkarni (1997), Application of isotopes to study karst

hydrogeology, In: Engineering Geology and the Environment (Ed; Marinos, Koukis, Tsiambaos, Stoumaras), Balkema, Rotterdam, pp. 1637-1622

20. S.M. Rao, K.M. Kulkarni (1997), Isotope hydrology studies on water resources in western Rajasthan, Current Science, Vol.72, No.1, pp. 55-61

21. S.V. Navada, A.R. Nair, S.M. Rao, B.L. Paliwall, C.S. Doshi (1993), Groundwater recharge studies in arid region of Jalore, Rajasthan using isotope techniques, Journal of Arid Environments, 24, pp.125 – 133

22. V. Sridhar, S.S. Merh, J.N. Malik (1999), Late Quaternary drainage distribution in Northwest India: A geoarchaeological enigma, In; Vedic Sarasvati, Memoirs Geological Society of India, No. 42, pp187 – 204

23. A.S. Rajawat, C.V.S. Sastry, A. Narain (1999), Application of pyramidal processing of high resolution IRS 1-C data for tracing migration of Sarasvati River in parts of Thar Desert, ibid, pp 259 – 272

24. S.M. Rao (2003), Use of isotopes in search of lost river, Journal of Radioanalytical and Nuclear Chemistry, Vol. 257, No.1, pp 5 – 9

25. M.A. Geyh, D. Ploethner (1995), An applied palaeohydrological study in Cholistan, Thar Desert, Pakistan, In: Application of Tracers in Arid Zone Hydrology (Ed: Eilon M. Adar, Christian Leibundgut), Proc. Vienna Symp. Aug. 1994, IAHS Publication no.232, pp. 119 – 127

26. A.E. Ramalingam (1980), Studies on seawater intrusion in a coastal aquifer north of Madras in Tamil Nadu. Proc. 48th Research Session of the CBIP, Hyderabad, Vol.2. pp. 133

27. K. Shivanna, S.V. Navada, A.R. Nair, S.M. Rao (1993) Isotope and geochemical evidence of past seawater salinity in Midnapore ground waters, Isotope Techniques in the Study of Past and Current Environmental Changes in the Hydrosphere and Atmosphere, Proc. Intl. Symp. IAEA, Vienna, pp. 199 – 212

28. K.M. Kulkarni, S.V. Navada, A.R. Nair, S.M Rao, K.Shivanna, U.K. Sinha, S. Sharma (1997), Drinking water salinity problem in coastal Orissa, India – Identification of past transgressions of sea water by isotope investigation, Isotope Techniques in the Study of Environmental Change, Proc. Intl. Symp. IAEA, Vienna, pp. 293 - 306

29. DANIDA (1993) Orissa Drinking Water Project – Hydrogeological Investigations, Danish International Development Assistance Report, 8 Vols. Royal Danish Embassy, New Delhi

30. S. Sharma, K.M. Kulkarni, U.P. Kulkarni, A.S. Deodhar, S.V. Navada, P.K. Jain (2003), Isotope Hydrogeochemistry of Groundwater in Purna River

Basin, Maharashtra, India, In: Book of Extended Synopses, International Symposium on Isotope Hydrology and Integrated Water Resources Management, IAEA, Vienna, pp61-62

31. W.T. Holser, M. Magaritz, J. Wright (1986), Chemical and isotopic variations in the world ocean during Phanerozoic time, In: Lecture Notes in Earth Sciences, Vol.8, (Ed: O. Walliser), Global Bio-Events, Springer, Heidelberg, pp 63 -74

32. K. Shivanna, U.K. Sinha, T.B. Joseph, S. Sharma, S.V. Navada (2000), Isotope hydrological investigation in arsenic infested areas of West Bengal, Proc. International Conference on Integrated Water Resources Management for Sustainable Development (ICIWRM), New Delhi, pp 490 - 500

33. P.K. Aggarwal, A.R. Basu (2000), Isotope Hydrology of Groundwater in Bangladesh: Implications for characterization Migration of Arsenic in Groundwater, Preliminary Report of Investigations, IAEA, Vienna

34. N.Ahmed, P.K. Aggarwal, K.M. Kulkarni, A.R. Basu, M. Ali, S.A. Tarafdar, A. Hussain (2003), Use of environmental isotopes to study the recharge mechanisms and arsenic pollution of Bangladesh groundwater, In: Book of Extended Synopses, International Symposium on Isotope Hydrology and Integrated Water Resources Management, IAEA, Vienna, pp 6 - 8

35. B.Th. Verhagen (2003), Isotope Hydrology and its impact in the developing world, Journal of Radioanalytical and Nuclear Chemistry, Vol.257, No.1, pp 17 - 26

36. K.Tirumalesh, K. Shivanna, J. Noble, K.K. Narayan (2006) Nuclear Techniques to investigate source, origin of groundwater pollutants and their flow path at the Indian Rare Earth Ltd., Cochin, Kerala, Proc. Int.Conf. on Application of Radiotracers in Chemical, Environmental and Biological Sciences (ARCEBS 06), Kolkata, Volume 2, pp 146 - 147

37. K.P. Seiler (2001), The origin of saline groundwaters in the area of the Valencia Lake, Venezuela, In: Man's Impact on Groundwater Systems, IHP-V, Technical Documents in Hydrology, No.39, Vol.V, UNECO, Paris, pp. 32-39

38. K.M. Kulkarni, S.V. Navada, S.M. Rao, A.R. Nair, U.P. Kulkarni, Suman Sharma (1996), Effect of the Holocene climate on composition of groundwater in parts of Haryana, India – Isotopic evidence, Isotopes in Water Resources Management, Proc. IAEA Symposium, Vienna, pp. 439 – 454

39. International Atomic Energy Agency (1992), Statistical treatment Data on Environmental Isotopes in Precipitation, Technical Reports Series No. 331, IAEA, Vienna

BIBLIOGRAPHY

1. Environmental isotopes in the hydrological cycle, Principles and applications (Ed. W.G. Mook), Introduction, IHP-V, Technical Documents in Hydrology, No.39, Vol.IV (Groundwater)& V (Man's Impact on groundwater systems), Unesco, Paris, 2000

2. Use of artificial tracers in hydrology, IAEA-TECDOC-601, IAEA, Vienna, 1991

3. Vedic Sarasvati – Evolutionary History of a Lost River of Northwestern India (Ed: B.P. Radhakrishna, S.S. Merh), Memoir 42, Geological Society of India, 2002

5

NEED FOR ENHANCED APPLICATION OF ISOTOPE HYDROLOGY & HOW TO ACHIEVE IT?

5.1 The Need

In spite of the considerable experience and quite a few facilities available in the country, the growth of application of isotope techniques is not commensurate with India's size, its climatic and hydrogeological diversity and the sheer range of its water resources problems.

The situation may not be unique to India and other developing countries. Australia also appears to be facing a similar problem. A recent publication [1] says 'the limited uptake of the method may be due to a number of factors. Until relatively recently, isotope hydrology has remained a largely academic pursuit, or measurements were the domain of national atomic energy agencies. A large fraction of information lies buried in the grey literature or in arcane technical journals. Interaction with local or regional water resources agencies has been the exception rather than the rule'

India has probably a stronger case for a more effective use of isotope techniques. A large portion of groundwater occurs in hard rocks in India. Normal hydrogeological and geophysical methods cannot fully describe such groundwater regimes to effectively plan their exploitation. It is essential to apply tracer hydrology methods, especially isotope geochemical techniques, for a comprehensive understanding of the systems. The examples of Manikaran geothermal springs, seepage studies in Salal tunnel and in some dams, given in the earlier chapters come under this category. Studies on percolation tanks in basaltic terrain and on recharge processes need isotope support.

It will not be long before we would be planning to exploit very deep ground waters (>500 m), both in the hard rock terrains of peninsular India as well as in the Ganga plains to meet our ever growing need for water. We would need good understanding of the recharge processes to these deep ground waters. Isotopes will have a large role in such investigations.

Arid zone hydrology is yet another area where the potential of isotope techniques has been effectively demonstrated for the characterization of the recharge processes, both palaeo and modern, and including the role of ancient rivers.

Interlinking of river basins

There has been an active revival of interest [2] in discussing possible interlinking of major river basins in India to ensure equitable distribution of otherwise grossly uneven distribution of rainfall, both spatially and temporally. This project, more gigantic than any envisaged earlier (one budget estimate: Rs.5600 billion) poses enormous challenges to Indian science and technology. We may have the necessary technology, but it is the recognition of the level of scientific inputs needed that needs to be ascertained.

The discussion on interlinking so far, appears largely to focus on the geographical feasibility with minimum lift, possible benefits to irrigation and drinking water supply, economic and legal considerations and on the possibility of inadvertent introduction alien flora in some cases.

It is important to keep in mind that the different components of the hydrological cycle such as surface water, soil moisture, shallow and deep ground waters are all interdependent. Any human activity on one component is likely to have either a beneficial or deleterious effect on the other components. Interlinking of river basins is likely to substantially influence the groundwater regimes all over the country. We should take this into account in the planning stage and while monitoring ecological impact during operation.

Any study on the effect of river basin interlinking on the groundwater regimes needs a complete database involving geological, geophysical, hydrogeological, and geochemical parameters of the groundwater regimes likely to be associated with the project besides river water parameters like discharges and chemistry. Hydrochemistry should include not only the major and minor ions, but also the trace ions like iodide, strontium, boron etc, which help in characterizing groundwater residence and dynamics. Good estimates of the present day river – groundwater interactions should be available to see when and where a given river is influent or effluent in its course and how the situation will be in the post-linking stage. To the above database list, we must add isotopes to have a complete picture and a better hold on planning and monitoring.

This itself is a huge task needing not only substantial scientific manpower, but also analytical laboratories. Interlinking is only a part of the country's programme of water resources development and management. The first task is human resources development.

5.2 How to achieve the Manpower for a More Effective Use of Isotope Hydrology?

It is only recently hydrologists and hydrogeologists are being increasingly associated with water resources development and management at the field level, which has been mainly a civil engineering activity. Some universities have hydrology courses at the post-graduate level. This is a good beginning and a foundation on which enhanced field application of isotope geochemistry can be planned.

Under the auspices of the Department of Atomic Energy at the national level and the International Atomic Energy Agency at the international level a number of training courses were conducted on the role of isotope hydrology in water resources development and management. The Centre for Water Resources Development and Management (CWRDM), Kozhikode has also carried out a few courses with the support of the Department of Science and Technology. The impact of these courses is not discernible to the extent envisaged. This is probably because those attending the courses are usually not the decision makers.

What appears essential is to start teaching isotope techniques to practising senior hydrologists/hydrogeologists (who are decision makers) through refresher courses as well as to hydrology/hydrogeology students at the post-graduate level in the universities. This teaching should be woven into a course on geochemistry for groundwater dynamics, not merely for water quality purposes. It is only natural that a student sees merit of any methodology, if taught in a formal course than in a more casual training activity. Such a course should enable the hydrologist to identify problems amenable to isotope hydrochemical solutions and approach specialist institutions for the necessary service.

A university course on isotope hydrology is of greater relevance to India than to developed western countries, not only because of our greater dependence on groundwater, but also because the methods are better suited to our tropical climate and its diversity. Isotope signatures are more distinct in our type of hydrological environment. Literature is replete with western universities doing their fieldwork in South America, Africa and the Middle East, but hardly in their own environment. Indian scientists, on the other hand, have tailor-made situations at home for applications of isotope hydrochemical methods for water resources development and management.

Any new university course depends on the availability of teachers, textbooks and laboratory facilities. The institutions with isotope hydrology facilities should be approached for providing teachers at the early stages and also help train university teachers in the field of isotope hydrology. As of now, those institutions are the Bhabha Atomic Research Centre, Mumbai, Physical Research Laboratory, Ahmedabad, National Geophysical Research Institute, Hyderabad, National Institute of Hydrology, Roorkee, Centre for Water Resources Development and Management, Kozhikode and Indian Agricultural Research Institute, Delhi. Indian Institute of Technology, Kharagpur is also reported to be in the process of setting up an isotope hydrology laboratory. These institutions should also be approached for laboratory facilities to arrange demonstration of measurement techniques to students till such time the universities themselves acquire the equipment. Since the objective of the course is to enable the student to identify problems amenable to isotope techniques, lack of in-house facilities should not be a great hindrance. The equipment becomes essential when the university starts research programmes in isotope hydrology.

Lecture notes should form the bulk of reading material at the early stages of the university programme with the possibility of such lecture notes evolving into textbooks at a later date. Besides the present book, there are already some books available for reference and help in preparing lecture material. They are

1. Guidebook on Nuclear techniques in Hydrology (IAEA, Vienna) 1983
2. Handbook of Environmental Isotope Geochemistry (Vol. 1&2) (Ed: P.Fritz & J.Ch. Fontes), Elsevier 1986
3. Chemical & Isotopic Groundwater Hydrology: The Applied Approach by E. Mazor, Marcel-Decker, 1997
4. Environmental Isotopes in Hydrogeology by I. Clark & P. Fritz, Lewis Publications, 1997
5. Isotope Tracers in Catchment Hydrology (Ed: Caroll Kendall, Jeffery Mc Donnel), Elsevier 1008
6. Environmental isotopes in the hydrological cycle – Principles and applications (6 volumes, Ed: W.G. Mook), Vol. I: Theory, Methods, Review, Vol. II: Atmospheric Water, Vol. III: Surface Water, Vol. IV: Groundwater: Saturated and Unsaturated Zone, Vol. V: Man's Impact on Groundwater Systems and Vol. VI: Modelling (IHP-V, technical Documents in Hydrology, No.39, Vol. I-VI, UNESCO, Paris, 2000)

Possible Syllabus

Lectures (15 - 20) covering

1. Introduction to tracer hydrology; concept of tracers, relevance to hydrological systems, examples of hydrochemical tracers
2. Introduction to isotope physics, isotopes of the water molecule and of dissolved salts (^{2}H, ^{3}H, ^{18}O, ^{13}C, ^{14}C and ^{34}S), their nomenclature and measurement, effect of condensation, evaporation and exchange processes on the isotope composition of the water molecule and dissolved salts.
3. Occurrence of environmental radioisotopes like tritium, carbon-14 and chlorine-36 in the hydrosphere, Dating of groundwater
4. Measurement of stable and radioactive isotopes in water samples including those of dissolved salts

5. Production and application of injected radioisotope tracers in short range studies. This would include selection, injection, assay and interpretation as well as the radiation safety aspects
6. Applications in surface water hydrology; stream flow measurement, stream flow component separation, seepage studies in canals, tunnels, dams and reservoirs, lake dynamics
7. Introduction to mathematical models to interpret isotope data in hydrology
8. Applications to studies on soil moisture movement in the unsaturated zone
9. Application of environmental isotopes in studies on recharge processes and interconnection between surface and groundwater bodies
10. Studies on groundwater salinization including seawater intrusion into coastal aquifers using environmental isotopes and chemistry
11. Applications in arid zone hydrology with special reference to palaeo waters and palaeo climates
12. Applications to studies on groundwater pollution, both from natural and anthropogenic sources.

Laboratory (1 week)

1. Familiarization with laboratory and field nucleonic equipment and their use for the assay of radioisotopes (^{82}Br, ^{131}I, ^{3}H etc.) in water samples
2. Mass spectrometric measurement of ^{2}H and ^{18}O in water samples and of ^{13}C and ^{34}S in dissolved carbonates and sulphates respectively
3. Sample preparation and use of liquid scintillation spectrometer for measurement of environmental tritium in water samples and of carbon-14 in dissolved carbonates.

Field-work (1 - 2 weeks)

Participation in planning a field study: This could include bibliography and collection of available conventional data, sample collection for isotope and chemical analyses, laboratory analyses and interpretation of all data.

The above is an example of a possible introduction to isotope hydrology as part of a hydrochemistry course.

References

1. A.L. Herczeg, A.J. Love, G.A. Harrington (2003), Environmental isotopes as groundwater management tools: Indispensable or superfluous, In: Book of Extended Synopses, International Symposium on Isotope Hydrology and Integrated Water Resources Management, IAEA, Vienna

2. Interlinking of Rivers in India: Retrospect and Prospect, Special volume, Journal of Applied Hydrology, Vol. XVI, No. 4A, October 2003.

Index

D

E

F

G

H

T

U

V

W

FSC
www.fsc.org
MIX
Papier | Fördert
gute Waldnutzung
FSC® C083411

Zeitfracht Medien GmbH
Ferdinand-Jühlke-Straße 7
99095 Erfurt, Deutschland
produktsicherheit@kolibri360.de